An Intuitive Introduction
to Complex Analysis
volume 2 (draft version 1)

Conformal
Mapping
and Its
Applications

Thomas J. Osler
Sky Pelletier Waterpeace

Note on Volume II, draft version:

This is a draft manuscript version of this text. It is intended to make the work available to benefit readers while the final production version is being prepared.

An Intuitive Introduction to Complex Analysis was originally intended to be published as a single volume. *Volume II: Conformal Mapping* was designed to be Chapter 9 of the original single-volume work. Preparing the manuscript for publication, the authors agreed that the work of Chapter 9 should be split off and published separately as *Volume II*. This conclusion was reached for two reasons: first, Chapter 9 had always been something of an oddity whose length comprised a significant proportion of the length of the other 8 chapters combined. Secondly, the material of Chapter 9 really stands on its own and makes for an effective treatment of conformal mappings.

In this draft version of *Volume II*, the original manuscript has been made available to allow students and instructors to benefit from it during the time until the final production-quality version is available. However, there are certain trade-offs to this approach, notably in the labeling of sections, pages, and figures. Since this Volume was originally Chapter 9 of a larger work, all numberings follow the convention from that first work; one easily-observed result is a preponderance of 9's prefixing labels, which are out of context in this publication. However, the authors concluded that the effort involved to amend this relatively minor incongruence would be better put to use in furthering the preparation of the final production-quality version of the text.

Hence, the authors request the indulgence of their dear readers in over-looking numbering and other such incongruities that arise from the unamended transition from included Chapter to standalone Volume. However, in the event that the authors have overlooked any significant content gaps arising from this transition, the reader is cordially asked to notify Sky Pelletier Waterpeace via email to waterpeace@rowan.edu, to make note of such omissions.

It is with great pleasure that this work is made available, with the hopes that it may be of service to students and instructors.

Part 1

9.1 Introduction   9.1
9.2 Meaning of the Laplacian operator   9.2
9.3 Stream lines in fluid flow   9.8
9.4 Velocity potential in fluid flow   9.16

Part 2

9.5 Conformal mapping of fluid patterns   9.26
9.6 The Jowkowski Transformation   9.31

Part 3

9.7 The electronic field   9.41
9.8 Heat flow   9.59

Part 4

9.9 The bilinear transformation   9.71
9.10 More properties of the bilinear transformation   9.88

Part 5

9.11 The Poisson integral formula   9.96

Part 6

9,12 The Shwarz Chistoffel Transformation   9.114

Appendix 1 Solutions to Problems   P 9.1 to 9.14

Supplementary Problems   SP 9.1 to SP
9.32

CHAPTER 9

CONFORMAL MAPPING AND ITS APPLICATIONS

9.1 Introduction

In Chapter 2 and Chapter 5, we studied certain mapping properties of

analytic functions. In this chapter, we continue the study of the maps

effected by analytic functions, and show how they are applied to the study

of fluid flow, electrostatics, and steady state temperatures.

Recall that in Chapter 5 we saw that when $w = f(z) = u(x, y) + i\, v(x, y)$

is an analytic function then :

(a) Both u and v satisfy Laplace's equation:

$$u_{xx} + u_{yy} = 0 \text{ ,}$$

$$v_{xx} + v_{yy} = 0 \text{ .}$$

We say that u and v are harmonic functions or potential functions.

(b) The mapping effected by $w = f(z)$ preserves angles and their

sense (its a conformal mapping) at all points where $f'(z) \neq 0$.

(c) The curves $u(x, y) = c_1$ and $v(x, y) = c_2$ (where c_1 and

c_2 are constants), intersect at right angles at points where

$f'(z) \neq 0$.

· The reader should review sections 5.4 and 5.5 before continuing

if these items have been forgotten.

9.2 Meaning of the Laplacian operator

Every physical problem involves three dimensional space and time.

This means that the three real space variables x, y and z of Cartesian

coordinates and time t are encountered.

Imagine now a problem of fluid flow.

Suppose we are in a large channel of

water and that the water is flowing at

constant velocity in the direction of

the y - axis. Suppose also that a very long cylindrical pipe is placed in

the position of the z - axis. We would expect the stream lines of the fluid

to resemble those shown in the figure. A stream line is the actual path

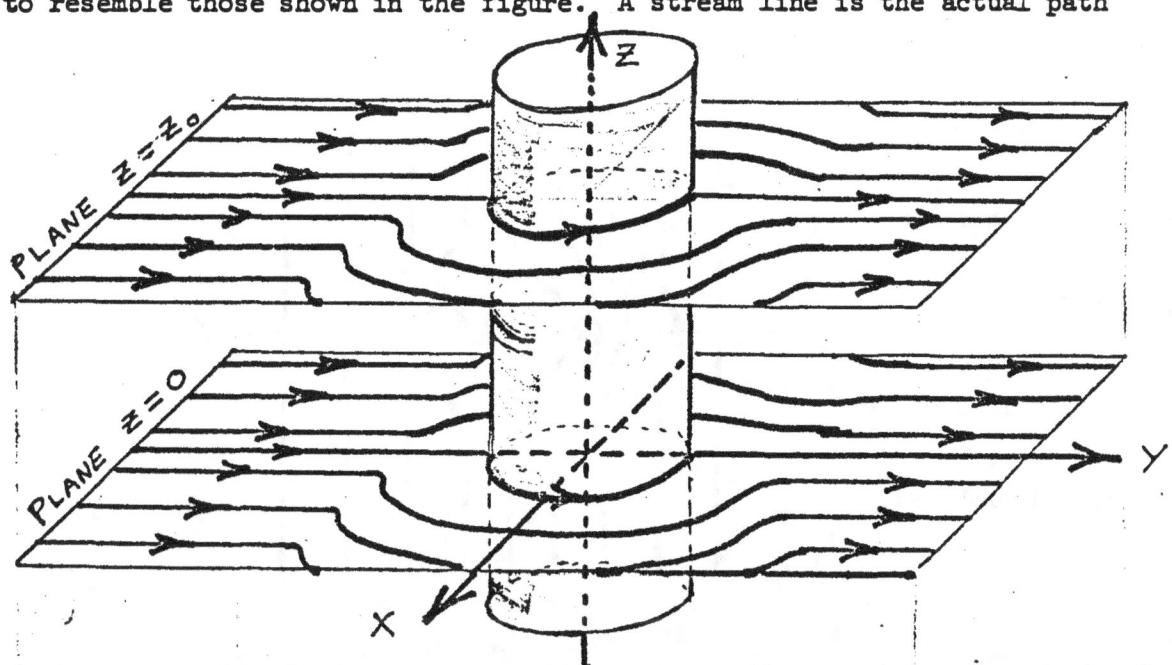

taken by a particle of the fluid. Suppose that the velocity of the fluid in the channel stays the same for a very long time at each point. During this time we say that we are in the steady state, and the mathematical description of the motion will not involve the variable t . Also, suppose the pipe is very long, and that the surface of the fluid and the walls of the channel are very far from the origin of coordinates. In this case, we can simplify our mathematical model by assuming that the water extends to infinity in every direction and that the pipe is infinite in length. In this case, the stream lines in the plane z = 0 look exactly like the stream lines in any other plane $z = z_0$. This means that the mathematical description of the fluid flow will not involve the variable z . We say that the motion involves only two space dimensions. We need only look now at the x, y plane to see the description of the stream lines, and their equation

should take the form $\psi(x, y) = c$,

where c is a parameter which is con-

stant on any stream line.

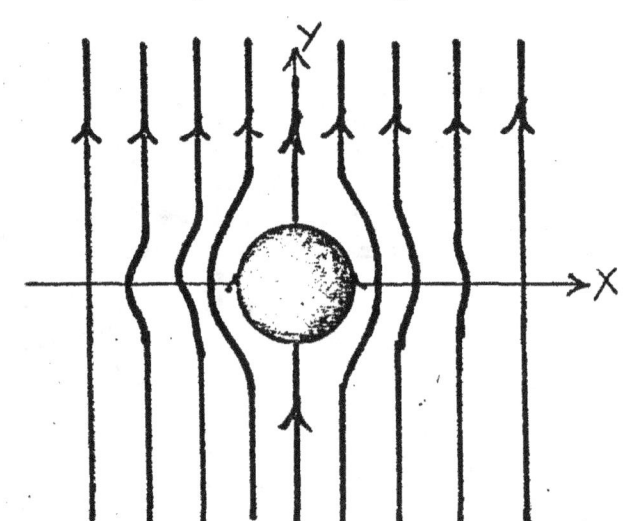

In the remainder of this chapter,

we will deal with physical problems

involving two space dimensions (x and y variables) and in the steady state (no variation with time). When a physical phenomenon is in the steady state, some type of balance or equilibrium has been obtained. Were this not so, the lack of balance would cause the system to alter, thus destroying the steady state. The equation which describes the equilibrium condition of a great many physical problems is the equation of Laplace

$$(1) \quad \nabla^2 u(x, y) = \frac{\partial^2 u(x, y)}{\partial x^2} + \frac{\partial^2 u(x, y)}{\partial y^2} = 0$$

In (1) , u (x, y) is some function which describes the physical problem. It might be the temperature at the point (x, y) , for example. The operator $\nabla^2 = \frac{\partial^2}{\partial x^2} + \frac{\partial^2}{\partial y^2}$ is called the two dimensional Laplacian operator.

The meaning of the expression $\nabla^2 u$ in physical problems is contained in the following idea:

(2) | $\nabla^2 u$ at a point (x, y) is a measure of the difference between the value of u at (x, y) , and the average of the values of u in an infinitesimal neighborhood surrounding the point (x, y) .

Suppose that u (x, y) describes the temperature at each point (x, y) in some solid. If the temperature is in the steady state, then the temperature at the point (x, y) should equal the average of the temperatures at points immediately surrounding it. If the temperature at (x, y) was _lower_ than the average of the temperatures surrounding it, heat would flow into the point (x, y) and cause the temperature there to rise, thus destroying the steady state. This example describes why Laplace's equation $\nabla^2 u = 0$ is the _equilibrium_ or _steady state_ equation.

We will now suggest why (2) is true. Since we must investigate u near the point (x, y), it is reasonable to expand u in a Taylor's series about (x, y).

$$(3) \quad u(x+h, y+k) = \sum_{n=0}^{\infty} \sum_{m=0}^{\infty} \frac{\partial^{m+n} u(x,y)}{\partial x^m \partial y^n} \frac{h^m}{m!} \frac{k^n}{n!}$$

$$= u(x,y) + u_x(x,y) h + u_y(x,y) k$$

$$+ u_{xx} \frac{h^2}{2} + u_{xy} h k + u_{yy} \frac{k^2}{2}$$

$$+ u_{xxx} \frac{h^3}{6} + \ldots$$

Perhaps the reader is seeing (3) for the first time. Notice how it is a natural outgrowth of the familiar Taylor's series. The difference between u at (x, y) and at a nearby point $(x+h, y+k)$ is

(4) $\quad u(x + h, y + k) - u(x, y) = u_x \ h + u_y \ k$

$$+ u_{xx} \ \frac{h^2}{2} + u_{xy} \ h \ k + u_{yy} \ \frac{k^2}{2}$$

$$+ u_{xxx} \ \frac{h^3}{6} + \ldots$$

The average of these differences over the infinitesimal square sur-

rounding (x, y) of side length $2 \in$ is

(5) $\quad \dfrac{1}{4 \in^2} \displaystyle\int_{-\in}^{\in} \int_{-\in}^{\in} \left[u(x + h, y + k) - u(x, y) \right] dk \ dh$.

If we replace the integrand here by the right side of (4) and integrate

we see that: (a) The first term gives

$$\frac{u_x (x, y)}{4 \in^2} \int_{-\in}^{\in} \int_{-\in}^{\in} h \ d \ k \ d \ h = 0$$

since h is an odd function. (If you don't see this, then actually

perform the integration) .

(b) For the same reason, the second term

$$\frac{u_y (x, y)}{4 \in^2} \int_{-\in}^{\in} \int_{-\in}^{\in} k \ d \ k \ d \ h = 0$$

(c) The third term does not vanish since we have

$$\frac{u_{xx}}{4 \in^2} \int_{-\in}^{\in} \int_{-\in}^{\in} \frac{h^2}{2} \ d \ h \ d \ k =$$

$$\frac{u_{xx}}{2 \in} \left[\frac{h^3}{6} \right] \Bigg|_{-\in}^{\in} \qquad = \frac{u_{xx} \in^2}{6}$$

(d)   The fourth term vanishes in the same way as the first two terms.

(e)   The fifth term resembles (c) and is

(7) $$\frac{u_{yy}\,\epsilon^2}{6}$$

(f)   The remaining terms will involve ϵ to a higher power than two,

and thus they will be ignored.

Combining (5) , (6) and (7) we have

(8) $$\frac{1}{4\epsilon^2} \int_{-\epsilon}^{\epsilon} \int_{-\epsilon}^{\epsilon} \left[u(x+h,\,y+k) - u(x,\,y) \right] \, dk\, dh \approx \frac{\epsilon^2}{6} \nabla^2 u$$

This last relation is the mathematical statement of (2).

Thus we see that the two dimensional equation of the steady

state is Laplace's equation. Since both the real and the imaginary parts

of an analytic function

$$w = f\,(z) = u\,(x,\,y) + i\,v\,(x,\,y)$$

satisfy this equation

$$\nabla^2 u = 0 \,,$$
$$\nabla^2 v = 0 \,,$$

it is not surprising that analytic functions are very useful in describing

many physical problems which are at the steady state, and involve only two

space variables. We will explore some of these applications in the
following sections.

9.3 Stream lines in fluid flow

One important application of analytic functions is to the study known
as hydrodynamics, aerodynamics or fluid dynamics. Suppose we have a two
dimensional flow of fluid in the steady state which has the additional
properties:

    (a) The fluid is incompressible.

    (b) The fluid is non-viscous.

    (c) The velocity of the fluid is derivable from a potential.

The assumption that the fluid is incompressible implies that the density of
the fluid is constant. A viscous fluid would create a tangential frictional
force on an obstacle in its path. Since our fluid is non-viscous, any force
it creates on the obstacle must be perpendicular to the obstacle's surface.
The assumption that the velocity of our fluid can be obtained from a potential
(harmonic) function will be explained in the next section.

A fluid satisfying the above assumptions is called an ideal fluid. Such
a fluid never really exists in nature, but is often approximated. We will

see that analytic functions of a complex variable enable us to visualize these ideal fluids.

Since we have seen that the real and the imaginary parts of the analytic function $f(z) = u(x, y) + i\, v(x, y)$ satisfy the two dimensional equation of the steady state, it comes as no surprise that they are useful in examining the flow of an ideal fluid. Look at Figure 9.1 . Here the level lines $u(x, y) = $ constant and $v(x, y) = $ constant are sketched over the complex z – plane for eight different analytic functions $f(z) = u + i\, v$. The level lines for u are shown broken while the level lines for v are shown solid. Do not the solid lines v = constant in each sketch look like the stream lines of some particular fluid flow? We call the function $v(x, y)$ the stream function. The flow described by each of the eight diagrams in Figure 9.1 is outlined below:

Diagram (i) : A uniform horizontal flow from left to right.

Diagram (ii) : A flow around a right angled corner.

Diagram (iii) : Fluid emerges from the origin. We have a source at z = 0 .

Diagram (iv) : A vortex or whirlpool is at z = 0 .

Figure 9.1 Table of Stream Lines for Various Fluid Flows

Stream lines $v(x,y) = $ constant ——————

Velocity potential $u(x,y) = $ constant — — — — —

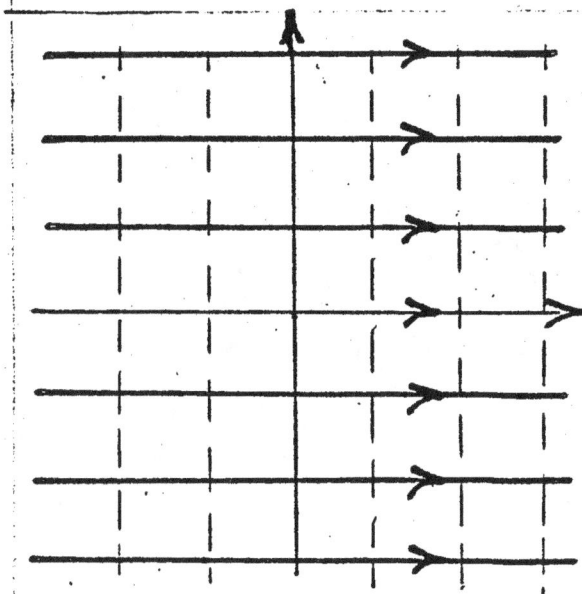

(i) Uniform horizontal flow.

$$w = z$$

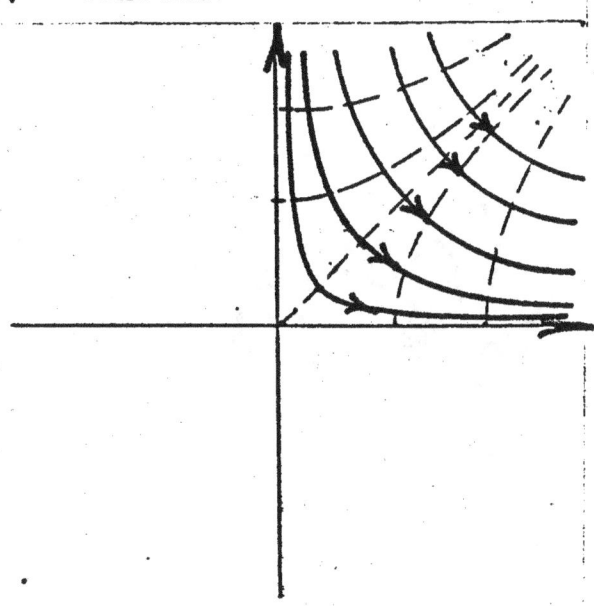

(ii) Flow around a corner.

$$w = z^2$$

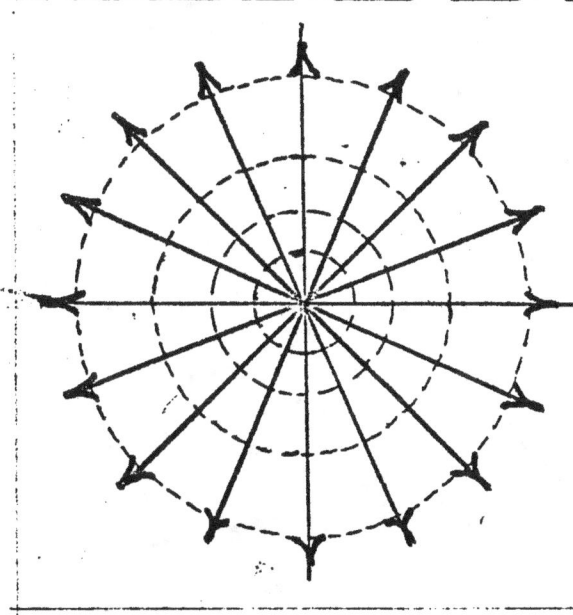

(iii) Source at $z = 0$.

$$w = \log z$$

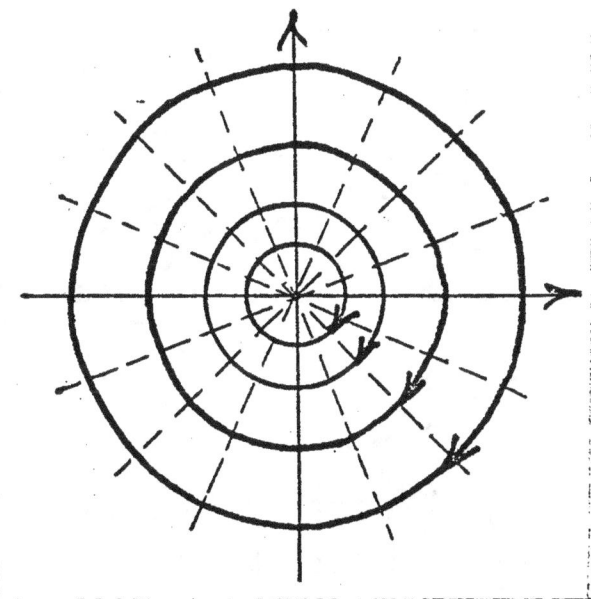

(iv) Vortex at $z=0$.

$$w = i \log z$$

Figure 9.1   Cøntinued

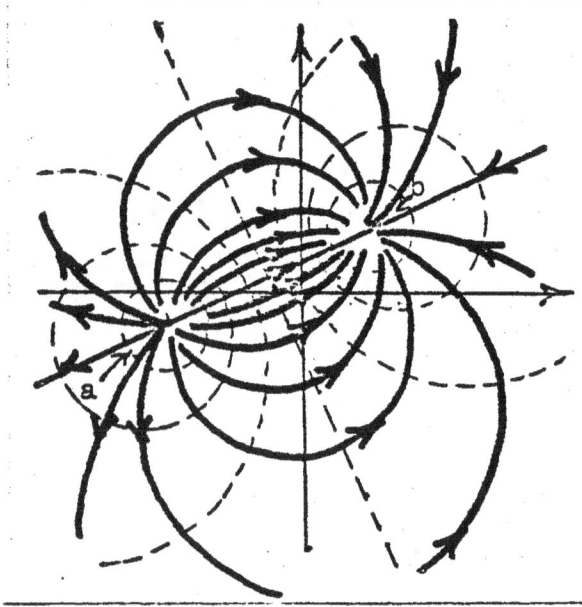

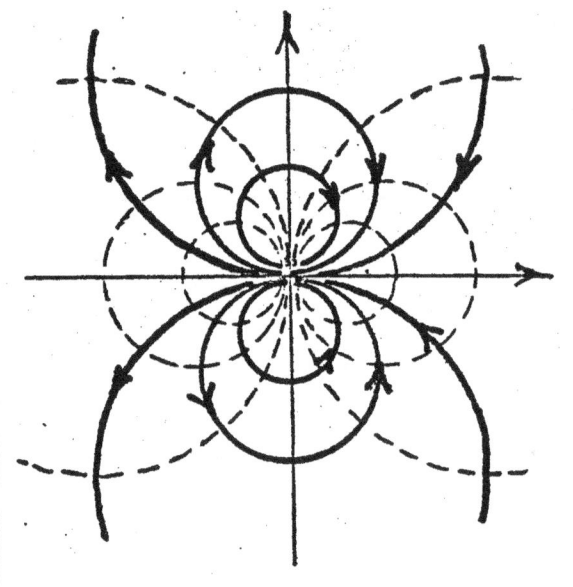

(v)   Source at $z = a$ and a
      sink at $z = b$.

$$w = \log \frac{z-a}{z-b}$$

(vi)   Dipole at $z = 0$.

$$w = \frac{1}{z}$$

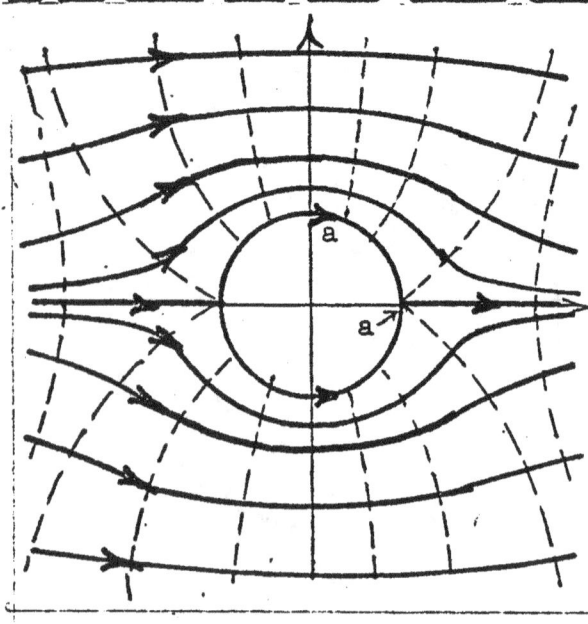

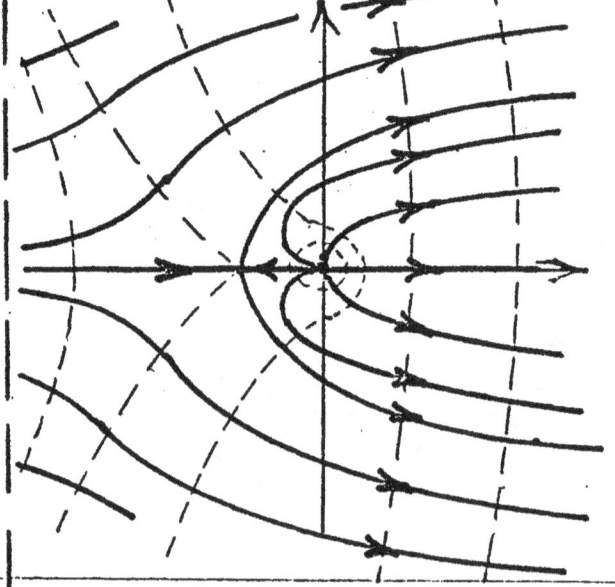

(vii)   Cylinder of radius a
       in a uniform flow .

$$w = z + \frac{a^2}{z}$$

(viii)   Source at $z=0$ in a uniform
horizontal flow from left to right.

$$w = Vz + m \log z$$

Diagram (v) : Fluid emerges at z = a and is absorbed at z = b.

We have a source at "a" and a sink at "b" .

Diagram (vi) : A dipole exists at the origin. A dipole consists

of a large source and sink separated by an infinitesimal distance.

Diagram (vii) : A cylindrical pipe of radius "a" is placed in

a uniform flow of fluid from left to right.

Diagram (viii) : A source of fluid is placed at z = 0 into a

uniform flow from left to right.

We will see in the next section how the direction of the fluid flow in

each diagram, as indicated by the arrow heads on the stream lines, is de-

termined.

We see that the imaginary part v (x, y) of each of the analytic functions

considered in Figure 9.1 defines a family of stream lines for the flow of a

perfect fluid. We will see in the next section how the corresponding real

part u (x, y) is used to determine the velocity of the fluid.

Example 1

Determine the equations of the stream lines shown in Figure 9.1 (ii).

Solution

Since $z^2 = (x + i\,y)^2 = x^2 - y^2 + 2\,x\,y\,i$, we see that the imaginary

part is $v\,(x,\,y) = 2\,x\,y$. Thus the equations of the stream lines are

$$x\,y = c$$

where c is a parameter which is constant along any stream line.

Example 2

Determine the stream function of the fluid flow in Figure 9.1 (vii).

Solution

We must find the imaginary part of the function $f\,(z) = z + a^2/z$. We

find $f\,(z) = x + i\,y + \dfrac{a^2}{x + i\,y}$

$$= x + i\,y + \frac{a^2\,(x - i\,y)}{x^2 + y^2}$$

(1) $$= x + \frac{a^2\,x}{x^2 + y^2} + i\left[y - \frac{a^2\,y}{x^2 + y^2}\right].$$

Thus the stream function is $v\,(x,\,y) = y - \dfrac{a^2\,y}{x^2 + y^2}$, and the family of

stream lines of fluid flow past a cylinder of radius a centered at the origin

is obtained by setting v equal to an arbitrary parameter.

Problems

Write the equations of the stream lines for the flows described by each

of the following:

3.1 The uniform flow in Figure 9.1 (i).

3.2 The flow described by the source at z = 0 in Figure 9.1 (iii).

3.3 The flow described by the dipole in Figure 9.1 (vi).

Any stream line can be thought of as the boundary along which the fluid must flow. This is possible because the assumption that the fluid is non-viscous means that no friction is created as the fluid passes over the boundary. As an example, we could consider one of the hyperbolas in Figure 9.1 (ii) as the boundary rather than the right angled corner.

Example 3

In the diagram we see two infinite plates at right angles to each other. Fluid continually enters the shaded region through the slit where the plates meet. When the steady state is reached, describe the steam lines.

Solution

Look at Figure 9.1 (iii) . If we use the two stream lines along the positive real and imaginary axies as the boundaries of our fluid, we see at once that the stream lines in the first quadrant of this figure satisfy the requirements of ~~our~~ problem. The stream lines are simply straight lines

emerging from the origin.

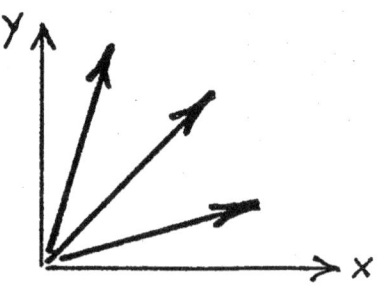

Problems

3.4 The edges of a metal boundary are shown in the diagram. Fluid

continually enters the shaded region

through the slit at A and leaves

at B at the same rate. Determine

the stream lines.

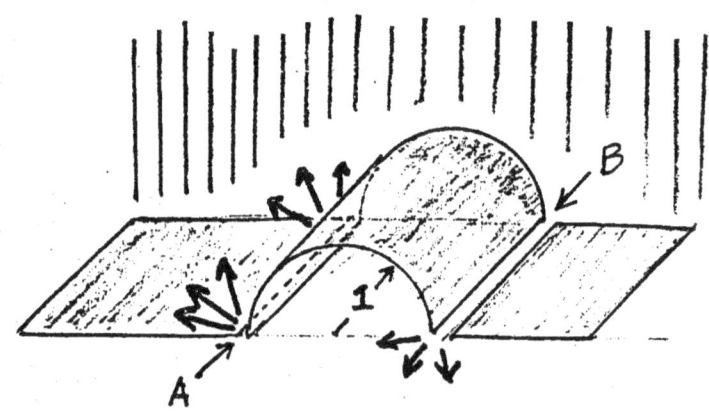

3.5 The semi-circular hump is placed on the bottom of a flat stream bed

where the fluid had been flowing

uniformly from left to right.

Determine the equations of the

stream lines.

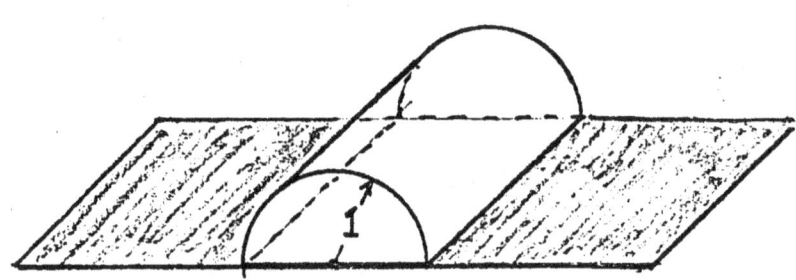

In some cases a stream line will divide and become two or more stream

lines at a particular point. Look at the stream line entering along the

negative real axis in Figure 9.1 (vii). At the point z = - a it splits

in two and travels round the cylinder to the point z = a where the two stream lines once again merge and continue to infinity along the positive

real axis. At each point, we can describe the velocity of the fluid by

means of a vector tangent to a stream line. Since there is no well de-

fined tangent at the points $z = \pm a$, the velocity of the fluid must

be zero there. We call these points <u>stagnation points</u> of the fluid. Look

again at the stagnation points in Figure 9.1 (vii) and observe that the

level lines for u and v do not intersect at right angles. This means

that the points $z = \pm a$ are points where the analytic function

$$w = f(z) = z + \frac{a^2}{z}$$

describing the flow fails to define a conformal mapping. Thus $f'(z) = 0$

when z is a stagnation point.

Example 4

Demonstrate analytically that the function

$$w = f(z) = z + \frac{a^2}{z}$$

describes a flow with stagnation points at $z = \pm a$. Show that the

stream line along the real axis splits at these points to form the circle

$\left| z \right| = a$ as shown in Figure 9.1 (vii).

Solution

The stagnation points occur where $f'(z)$ is zero. Thus

$$f'(z) = 1 - \frac{a^2}{z^2} = 0 \,,$$

and we have $z = \pm\ a$ as the stagnation points. Along the real axis,

$f(z)$ itself is real and thus $v(x,\ y) = 0$ is a stream line along the

real axis. In Example 2 we saw that

$$v(x,\ y) = y \left[1 - \frac{a^2}{x^2 + y^2} \right] \, .$$

Thus $v(x,\ y) = 0$ in two circumstances:

    (a) When $y = 0$ (real axis)

    (b) When $x^2 + y^2 = a^2$ (circle of radius a).

Problem

3.6 Find the location of the stagnation point of the flow in Figure 9.1

(viii) .

9.4 Velocity potential in fluid flow

The stream lines of fluid flow were found as the level lines of the

imaginary part of the appropriate analytic function in the previous section.

How can we find the velocity of the fluid at each point (x, y) ? The

velocity is a vector quantity which is tangent to the stream lines.

In a previous course, the student probably studied the gradient of the

function $u(x, y)$ denoted by ∇u and defined as a vector whose

horizontal component is u_x and whose vertical component is u_y . We

will denote this vector by a complex number

(1) $\nabla u = u_x + i\ u_y$.

The gradient (1) has the following properties:

      (a) ∇u at each point (x, y) is perpendicular to the level

      line $u = $ constant through this point.

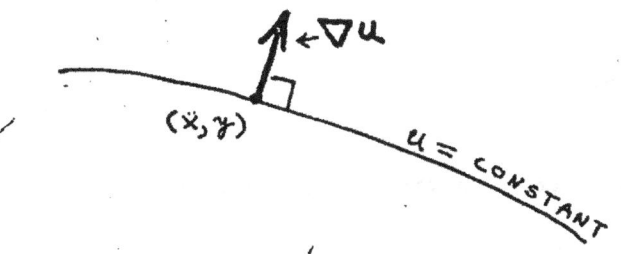

      (b) The modulus of the gradient, $|\nabla u| = \sqrt{u_x{}^2 + u_y{}^2}$ is

      the directional derivative $\dfrac{d\,u}{d\,s}$ in the direction of the most rapid increase in u . To understand this statement, think

of $u = u(x, y)$ as^a surface in (x, y, u) space. Then $\left| \nabla u \right|$ is

the slope we would experience if we were climbing up this surface

at the point (x, y, u) in the direction of greatest steepness.

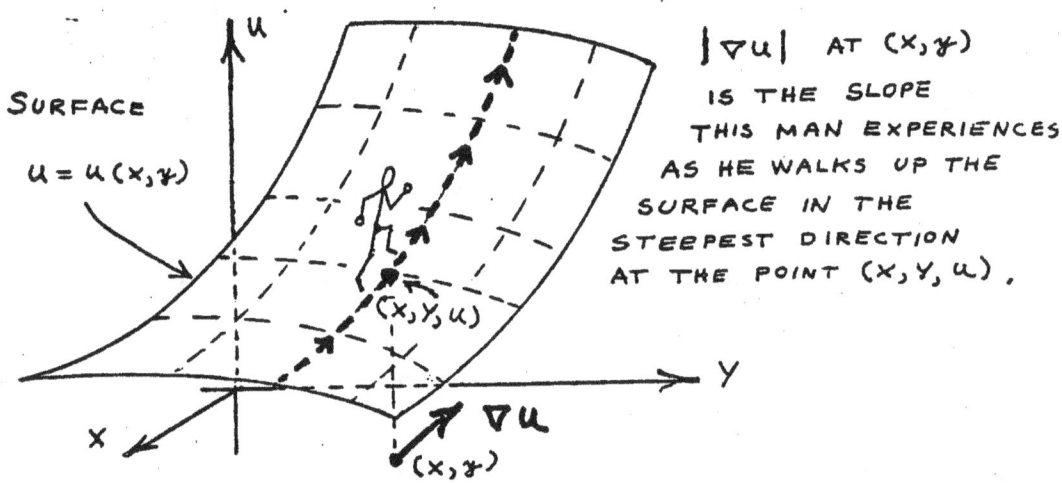

Now return to Figure 9.1 . Notice that at each point the level lines

for $u(x, y)$ are perpendicular to the stream lines. By property

(a) discussed above, ∇u must be tangent to the stream lines. Since

velocity vectors are also tangent to stream lines, could it be that ∇u

describes the velocity vector of our ideal fluid flow? Yes, this is the

case. We have "velocity vector" $= \nabla u = u_x + i\,u_y$, and since $u_y = -v_x$

(one of the Cauchy-Riemann equations from section 5.3) we have

(2) Velocity Vector $= u_x - i\,v_x$.

The derivative of the function $f(z) = u + i\,v$, $\dfrac{d\,f(z)}{d\,z}$, can be computed using dz as ∂x (recall section 5.1) and we have

(3) $\dfrac{df}{dz} = \dfrac{\partial f}{\partial x} = \dfrac{\partial(u + iv)}{\partial x}$

$\qquad = u_x + iv_x$

Comparing (2) and (3) we see that

(4) velocity vector $= \overline{f'(z)}$

Thus we see that an analytic function $f(z) = u + iv$ is a neat

compact package for the description of an ideal fluid with the following

features:

(a) The stream function is $v(x, y)$. The stream lines are

given by the equations $v(x, y) = c$, where c is an arbitrary

constant.

(b) The velocity potential is $u(x, y)$. Equipotential lines

are given by $u(x, y) = c$. The velocity vector itself is

$\overline{f'(z)}$, the complex conjugate of the derivative of $f(z)$.

We call $f(z)$ the complex potential of the fluid flow.

Example 1

Show that the function $f(z) = Vz$ is the complex potential of a uniform flow from left to right with speed V .

Solution

Since $f'(z) = V$, the velocity vector $\overline{f'(z)}$ is also V. The stream function is Vy and thus the stream lines are y = constant. The velocity potential is Vx and thus the equipotential lines are x = constant.

Example 2

Find (a) the complex potential, (b) the velocity vector and (c) the stream function when a cylindrical pipe of radius "a" is placed in a uniform horizontal flow with velocity V far from the pipe.

Solution

In Figure 9.1 (vii) we see a horizontal flow past a pipe of radius "a" described by the function

$$(5) \quad w = z + \frac{a^2}{z} .$$

The velocity of this flow is described by $\overline{w'} = 1 - \frac{a^2}{z^2}$.

When z is large (far from the pipe) the velocity is nearly 1. How can we adjust (5) so that it will have the same stream lines, but velocity V far from z = 0 ? If we multiply (5) by the constant V we will not alter the stream lines since the function V v(x, y) = constant and

$v(x,y)$ = constant   defines the same family of stream lines. Thus we

take

(6) $w = V(z + \dfrac{a^2}{z})$

as our new complex potential because the velocity $\overline{w'} = V - \dfrac{V a^2}{z^2}$

is nearly V far from the origin.

Example 3

Find the complex potential describing a uniform flow of velocity V

in the direction shown .

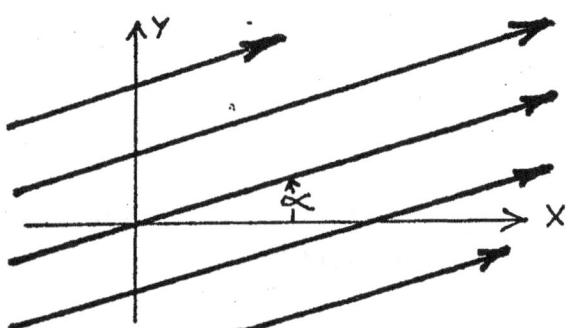

Solution

Since the velocity $V e^{i\alpha}$, is the complex conjugate of the derivative

of the complex potential we write

$$\overline{f'(z)} = V e^{i\alpha}$$

$$f'(z) = V e^{-i\alpha}$$

$$f(z) = V e^{-i\alpha} z .$$

Thus $V e^{-i\alpha} z$ is the desired complex potential.

Problems

4.1 Determine the velocity vector at each point (x, y) of a fluid flow described by the following complex potentials. Describe the stream lines.

(a) $w = m z^2$,   (b) $w = m \log z$,   (c) $w = m i \log z$,

(d) $w = m \log \dfrac{z - a}{z - b}$,   (e) $w = \dfrac{m}{z}$,   (f) $w = V z + m \log z$.

4.2 Find the points where the speed of the flow described in Example 2 is a maximum. Find this maximum speed.

Suppose we have a source of fluid at the origin with complex potential

described by $f(z) = m \log z$.
We call m the **strength** of the source.
What is the rate at which fluid

emerges from the origin? That

is,(assuming units of feet and

seconds)what is the number of

· foot of length perpendicular to the z plane.

cubic feet of fluid emerging per second? The figure shows a cylinder of

fluid at time t of   volume $= \pi r^2$

A short time later, $t + dt$, the radius of the cylinder increases to

$r + dr$. Since the velocity vector

has magnitude $\left| f'(z) \right| = \left| \dfrac{m}{z} \right| = \dfrac{m}{r}$,

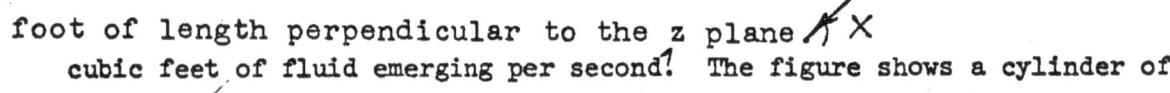

and since the speed is also described by $\frac{dr}{dt}$, we have $\frac{dr}{dt} = \frac{m}{r}$.

Thus the rate at which the volume increases is

$$\frac{d \text{ volume}}{dt} = \frac{d(\pi r^2)}{dt}$$

$$= 2 \pi r \frac{dr}{dt}$$

$$= 2 \pi r \frac{m}{r}$$

$$= 2 \pi m .$$

We have determined that:

The complex potential m log z describes a source of strength m at the origin in which fluid emerges at the rate 2π m units of volume per unit time for each unit of length perpendicular to the z plane.

Example 4

Fluid is being created at the point (-1, 0) at the rate of 3 cubic feet per second while fluid is absorbed at the rate of 2 cubic feet per second at the point (1, 0). Determine the complex potential.

Solution

If fluid were emerging from the origin at the rate of 3 cubic feet per second, the complex potential $\frac{3}{2\pi}$ log z would describe this source. To move the source to the point (-1, 0), we replace z by z -(-1) to get $\frac{3}{2\pi}$ log (z + 1). A sink at the origin in which fluid is lost at the rate of 2 cubic feet per second is described by $-\dfrac{2}{2\pi}$ log z . To translate this

sink to the point (1, 0) we replace z by $z-1$ to get $-\dfrac{2}{2\pi}$ log$(z-1)$.

Thus we have

$$f(z) = \frac{3}{2\pi} \log(z+1) - \frac{2}{2\pi} \log(z-1)$$

$$= \frac{1}{2\pi} \log \frac{(z+1)^3}{(z-1)^2}$$

as the complex potential for the flow.

Problem

$\mathbf{\circledcirc}$ 4.3 Fluid emerges from the point (0, 1) at the rate of 4π cubic feet per

second, while it is absorbed at the origin at the same rate. (a) Find the

complex potential. (b) Find the velocity vector. (c) Find the equations

of the stream lines and the equipotential lines.

Example 5

Discuss the flow described by $w = m \log(\sin z)$. Draw the stream lines

and equipotential lines.

Solution

Near the origin $\sin z \approx z$ and thus

$$w = m \log (\sin z)$$

$$\approx m \log z$$

for very small z . This is a source of strength   m   at the origin.

Near z = π , sin z ≈ π - z and we have

$$w = m \log (\sin z)$$

$$\approx m \log (\pi - z)$$

$$\approx m \log (-1)(z - \pi)$$

$$\approx m \log (z - \pi) + m \log (-1)$$

The term m log (-1) is simply a constant, and the addition of a constant

term to the complex potential in no way alters the physical flow being

described. The term m log (z - π) is a source of strength   m   at

z = π . Continuing in this way we find that sin z has zeros at all

integral multiples of π, and thus we have sources of strength   m   at

the points z = π n , n = 0 , $\pm$ 1, $\pm$ 2, $\cdots$. The stream lines of

$$w = m \log(\sin z)$$
$$= m \log |\sin z| + im \arg (\sin z)$$

are the curves arg (sin z) = c , where c is any constant. These curves

are seen in Figure 2.7 of Chapter 2. The equipotential lines are $|\sin z|$ =

constant and are seen in the same figure.

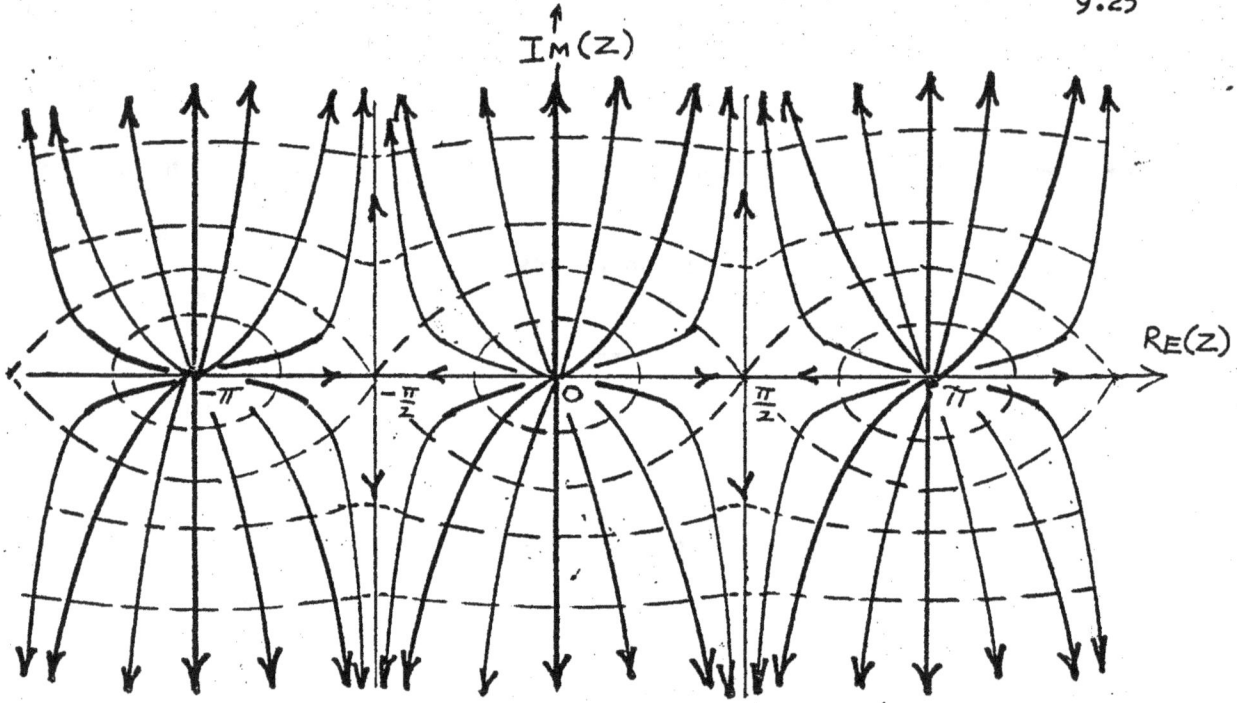

The Flow Pattern described by the complex potential

log (sin z) .   Re(log(sin z)) = c_1 - - - - - - - - - -

Im(log(sin z)) = c_2 ———————→

Problems

Discuss the flow described by each of the following complex potentials.

Locate all sources and sinks and determine their strength. Find all stagnation

points. Find the velocity vector and sketch the stream lines and the

equipotential lines.

4.4   w = log (cos z)

4.5   w = log (sinh π z)

4.6   w = log (csc z)

4.7   w = log (tan π z)

In this and the previous sections we looked at the level lines of the real and the imaginary parts of an analytic function $f(z) = u(x, y) + iv(x,y)$. We saw that the lines $v(x, y) = c$ resembled the stream lines of fluid flow and that the velocity vector could be obtained from the gradient of $u(x, y)$ as well as from $\overline{f'(z)}$. We did not derive these properties of the complex potential from basic assumptions on the nature of the fluid. We simply suggested through diagrams that the complex potential can provide a useful model of the flow of an ideal fluid. Careful derivations of these properties are given in texts on fluid dynamics.

9.5 Conformal mapping of flow patterns

In addition to using an analytic function f(z) as the complex potential

describing the flow of a fluid, we can also use mapping properties of

another analytic function to transform f(z) into a new flow pattern. To

be more precise, imagine that f(z) describes the flow shown, and that the

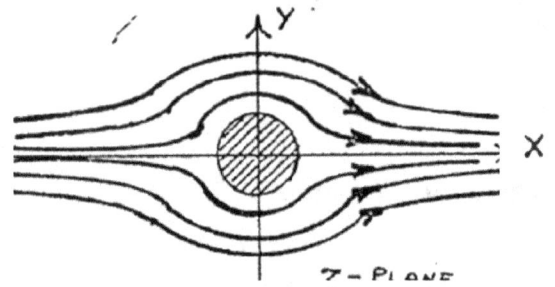

z - plane on which these

stream lines are drawn is

made of rubber.

Now consider another analytic function $\mathcal{S} = M(z)$, where $\mathcal{S} = s + i\,t$. This function $M(z)$ maps each point on the z - plane to a new point on the

$\mathcal{S}$ - plane. We can picture this mapping as a distortion of the rubber

surface to some new shape as shown.

The stream lines drawn on this

rubber sheet have also been

distorted. We can solve the

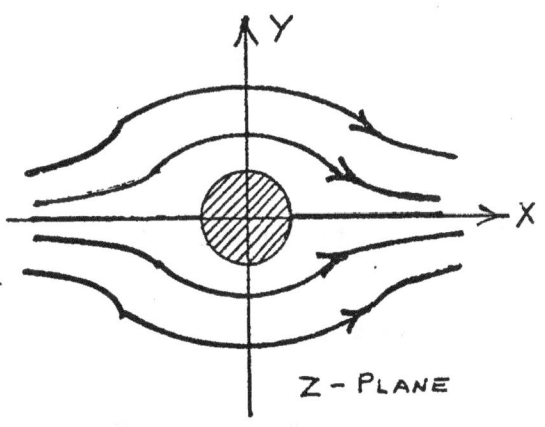

$\mathcal{S}$ - PLANE

relation $\mathcal{S} = M(z)$ for z in terms of $\mathcal{S}$ and denote the result by $z =$

$M^{-1}(\mathcal{S})$. Now the complex potential describing the new flow pattern is

$f(z) = f(M^{-1}(\mathcal{S}))$. Thus we see that a familiarity with mapping properties

of analytic functions $\mathcal{S} = M(z)$ can generate new models of fluid flow patterns.

Example 1

In Example 2 of the previous section we saw that the complex potential

$f(z) = V_0(z + 1/z)$ describes a

uniform horizontal flow which

encounters a cylindrical

obstacle of radius one at

the origin. Describe the

Z - PLANE

complex potential of a flow in the direction $\dfrac{\pi}{4}$ of velocity V far from the origin about a cylindrical obstacle of radius 5 with center at the

point 5 + 5 i .

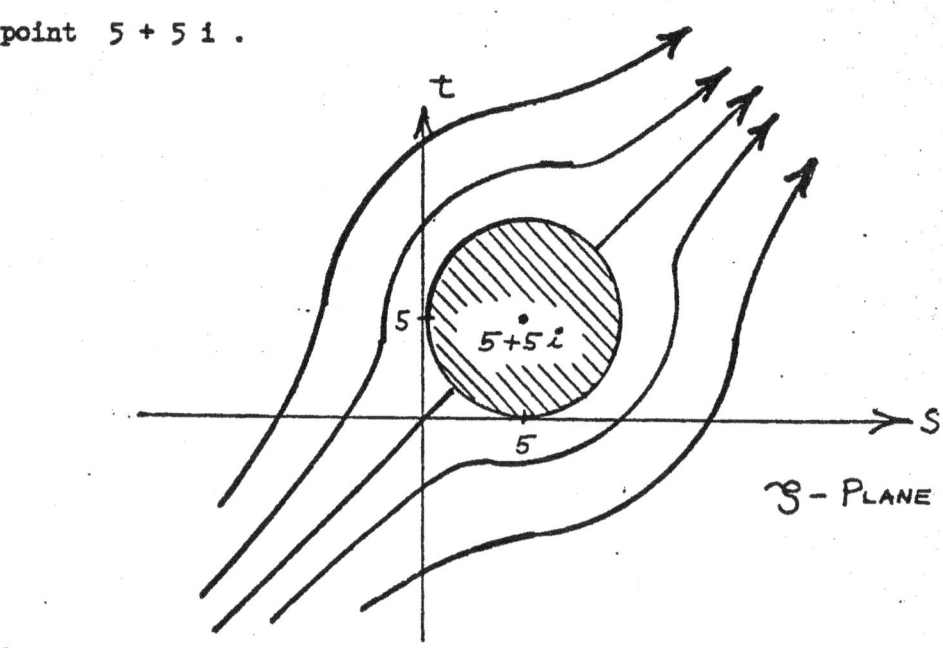

Solution

In section 5.5 we saw that the mapping function

$$ \mathcal{S} = M(z) = A\, e^{i\alpha}\, z + \mathcal{S}_0 $$

describes the following alterations of the complex z - plane:

        (i) A magnification by the factor A .

        (ii) A rotation through the angle α .

        (iii) A translation by the vector $\mathcal{S}_0$.

In our problem, we wish to:

        (i) Magnify the pattern in the z - plane by the factor

        5 (A = 5).

(ii) Rotate the pattern in the z - plane through

the angle $\frac{\pi}{4}$ $\left(\alpha = \frac{\pi}{4} \right)$.

(iii) Translate the center of the cylinder to the point

$5 + 5i \ (\zeta_0 = 5 + 5i)$.

Thus we see that the required mapping function is

$$\zeta = M(z) = 5 \ e^{i \ \pi/4} \ z + 5 + 5i \ .$$

Solving for z in terms of ζ we get

$$z = M^{-1} (\zeta)$$

$$= \frac{e^{-i \ \pi/4}}{5} \ \ (\zeta - 5 - 5i).$$

Substituting this result into the complex potential $f(z) = V_0 \ (z + 1/z)$

we get

$$f(z) = f(M^{-1}(\zeta)) =$$

(1)
$$F(\zeta) = V_0 \ \left(\frac{e^{-i \ \pi/4}}{5} \Big(\zeta - 5 - 5i \Big) + \frac{5}{e^{-i \ \pi/4} \Big(\zeta - 5 - 5i \Big)} \right)$$

We must now select V_0 so that the speed of the flow is V far from $\zeta = 0$.

For large $\left| \zeta \right|$, (1) is nearly

$$F(\zeta) \approx \frac{V_0 \ e^{-i \ \pi/4} \Big(\zeta - 5 - 5i \Big)}{5}$$

and thus the velocity is nearly

$$\overline{F'(\zeta)} \approx \frac{V_0 \, e^{i \, \pi/4}}{5} \quad .$$

Thus the speed V is $\dfrac{V_0}{5}$ and we select $V_0 = 5 \, V$. Finally our complex

potential is obtained from (1) as

$$V \left(e^{-i \, \pi/4} (\zeta - 25 - 25 \, i) + \frac{25}{e^{-i \, \pi/4} (\zeta - 25 - 25 \, i)} \right)$$

Problems

● 5.1   A cylindrical obstacle of radius 4 is placed in a uniform horizontal

flow of speed 5 from left to right. The center of the cylinder is at

$4 \, i$. Find the complex potential describing the fluid motion.

● 5.2   A cylindrical obstacle of radius 2 is placed in uniform vertical

flow (upward) of speed 10. The center of the cylinder is at $4 + 4 \, i$.

Find the complex potential describing the fluid motion.

The use of other analytic mapping functions $\zeta = M(z)$ produces further

flow patterns. It is important that the equipotential lines and the stream

lines intersect at right angles at points where the velocity is finite and

nonzero. Thus we want our analytic function $M(z)$ to be a conformal mapping

at all points in the interior of the region in the z - plane where the flow

occurs. We require that $M'(z) \neq 0$ in the interior of the region where we draw stream lines.

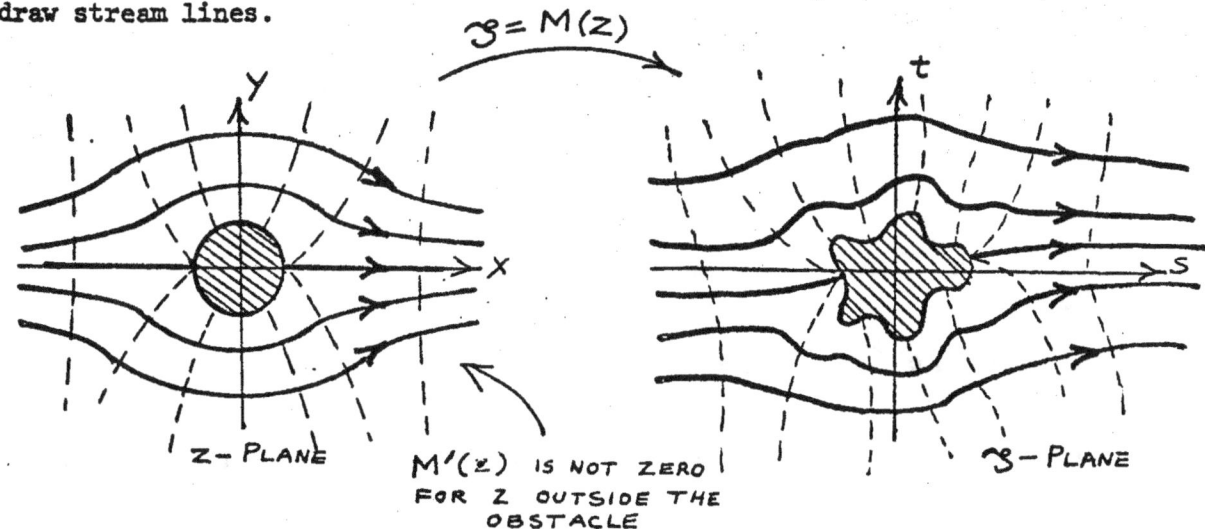

M'(z) IS NOT ZERO
FOR Z OUTSIDE THE
OBSTACLE

z - PLANE

$\mathfrak{z}$ - PLANE

9.6   The Joukowski transformation

One of the most important mapping functions is

(1) $\mathfrak{z} = M(z) = \tfrac{1}{2}(z + 1/z)$

known as the Joukowski transformation. Figure 9.2 shows the important

mapping properties of this function. The examples and problems of this

section illustrate their application.

Example 1

Verify the mapping illustrated in Figure 9.2 (i).

Solution

We must show that:

(i) The circle $z = e^{i\theta}$, $0 \le \theta \le 2\pi$ maps onto the line

segment $-1 \le s \le 1$, $t = 0$.

(ii) The Joukowski transormation (1) is conformal for

$|z| > 1$. This means that $M'(z)$ exists and is not

zero there.

If we set $z = e^{i\theta}$ in (1) we get

$$\Im = \tfrac{1}{2}(e^{i\theta} + e^{-i\theta}) = \cos\theta.$$

As θ varies from 0 to 2π, $\cos\theta$ varies from 1 to -1 and then back

to 1 along the real $\Im$ - axis. Thus item (i) has been demonstrated. From

(1) we have

$$M'(z) = \tfrac{1}{2}\left(1 - \frac{1}{z^2}\right).$$

Thus $M'(z)$ is undefined for $z = 0$, and is zero for $z = \pm 1$. Since

these points are not in $|z| > 1$, item (ii) has been demonstrated.

Remark:

We can tell that $M'(z) = 0$ for $z = \pm 1$ at first glance by looking

at Figure 9.2 (i). Since the curve in the z - plane at points A and C

is pinched at the corresponding points A' and C' in the $\Im$- plane we know

that $M(z)$ is not conformal (angle-preserving) at these points.

Example 2

Determine the complex potential describing the flow that results when a flat plate of width 2 is placed in a uniform vertical (upward) stream

of velocity V .

<u>Solution</u>

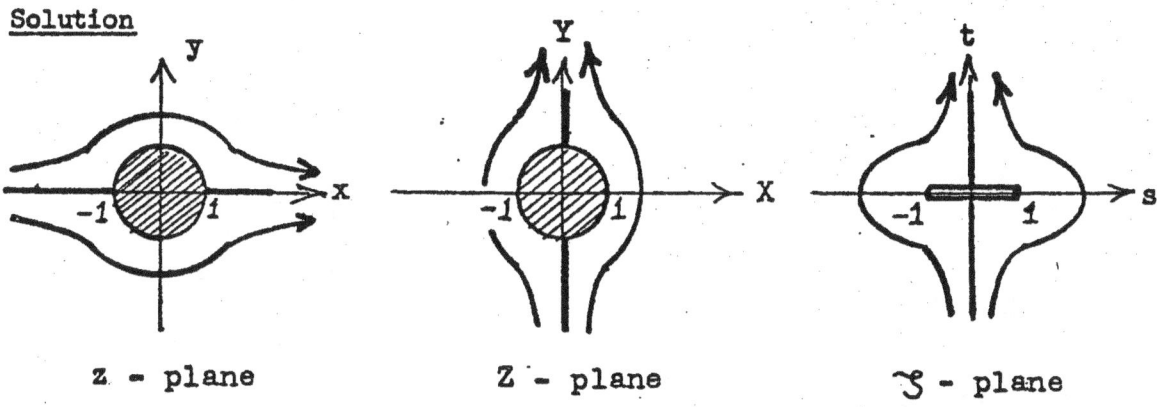

z - plane   Z - plane $\mathfrak{J}$ - plane

The flow on the z - plane is described by the complex potential

$$V_o(z + 1/z).$$

To map this pattern onto the Z - plane we must rotate the z - plane through

$90°$ by means of the mapping $Z = i z$.

Thus $z = - i Z$ and the complex potential becomes

$$V_o \left(- i Z + \frac{1}{- i Z} \right) =$$

(2) $-i V_o \left(Z - \frac{1}{Z} \right).$

From Figure 9.2 (i) we know that

(3) $\mathfrak{J} = \frac{1}{2}\left(Z + \frac{1}{Z} \right)$

will map the flow pattern on the Z - plane to that shown on the $\mathfrak{J}$- plane.

Solving (3) for Z in terms of $\mathfrak{J}$ we get

$$z^2 - 2\gamma z + 1 = 0$$

(4)
$$z = \gamma \pm \sqrt{\gamma^2 - 1}$$

We let $\sqrt{\gamma^2 - 1}$ denote that branch of the function that is continuous

on the γ - plane with a branch cut along the real axis from -1 to 1 and

such that $\sqrt{\gamma^2 - 1}$ is a positive

real number when $\gamma^2 - 1$ is a

positive real number.

We must take then

(5)
$$z = \gamma + \sqrt{\gamma^2 - 1} \quad,$$

branch is such that $\sqrt{1} = 1$

branch cut for $\sqrt{\gamma^2 - 1}$

for the minus sign in (4) would map the γ- plane onto the __interior__ of

$|z| = 1$ rather than the __exterior__ as is required. (See Figure 9.2 (ii)).

Substituting (5) into (2) we get

(6)
$$F(\gamma) = i V_0 \left(\gamma + \sqrt{\gamma^2 - 1} - \frac{1}{\gamma + \sqrt{\gamma^2 - 1}} \right)$$

as the complex potential. We must now select V_0 so that when $|\gamma|$ is large,

the velocity is $V i$. When $|\gamma|$ is large, $\sqrt{\gamma^2 - 1}$ is nearly γ and thus

(6) becomes

$$F(\gamma) \approx -2 i V_0 \gamma$$

The velocity is then, for large $|\mathcal{Z}|$,

$$\overline{F'(\mathcal{Z})} \approx \overline{-2iV_o} = 2V_o i$$

Thus $2V_o = V$ and (6) is at last

$$F(\mathcal{Z}) = -\frac{i}{2}\frac{V}{2}\left(\mathcal{Z} + \sqrt{\mathcal{Z}^2 - 1} - \frac{1}{\mathcal{Z} + \sqrt{\mathcal{Z}^2 - 1}}\right).$$

This is the required complex potential.

Example 3

A uniform flow of speed V in the direction $45°$ encounters a cylindrical obstacle whose cross section is the ellipse shown. Determine the complex potential describing the flow.

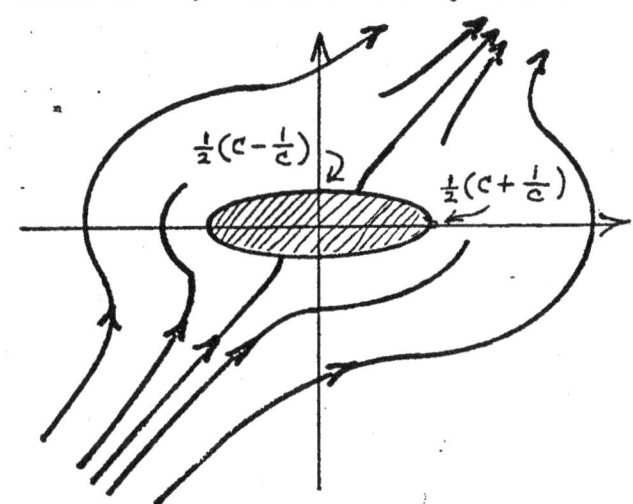

Solution

Examine the three diagrams shown.

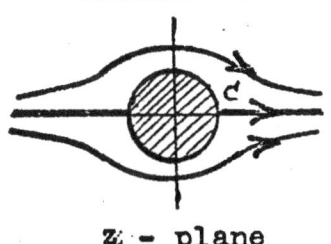
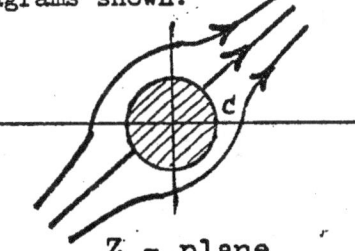
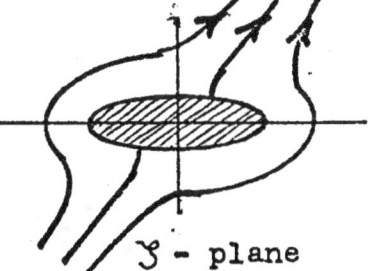

z - plane   Z - plane $\mathcal{Z}$ - plane

The complex potential describing the flow in the z - plane is

$$(7) \qquad V_o\left(z + \frac{c^2}{z}\right)$$

as was seen in Example 2, section 9.4 . To map the z – plane onto the Z – plane we require a rotation through $45°$. Thus $Z = e^{i\,\pi/4}\,z$ and $z = e^{-i\,\pi/4}\,Z$. Thus (7) becomes

(8)
$$V_o\left(e^{-i\,\pi/4}\,Z + \frac{c^2 e^{i\,\pi/4}}{Z}\right).$$

We can map the Z – plane onto the $\mathfrak{Z}$ – plane using the Joukowski transformation as shown in Figure 9.2 (iii). Thus

$$\mathfrak{Z} = \tfrac{1}{2}\left(Z + \frac{1}{Z}\right)$$

and

$$Z = \mathfrak{Z} + \sqrt{\mathfrak{Z}^2 - 1}$$

as was seen in (5) of Example 2 . Substituting this last relation into (8) we get

(9)
$$F(\mathfrak{Z}) = V_o\left[e^{-i\,\pi/4}\left(\mathfrak{Z} + \sqrt{\mathfrak{Z}^2 - 1}\right) + \frac{c^2\,e^{i\,\pi/4}}{\mathfrak{Z} + \sqrt{\mathfrak{Z}^2 - 1}}\right].$$

We must now select V_o so that when $|\mathfrak{Z}|$ is large, the velocity is $V\,e^{i\,\pi/4}$. For large $|\mathfrak{Z}|$, $\sqrt{\mathfrak{Z}^2 - 1} \approx \mathfrak{Z}$ and we have

$$F(\mathfrak{Z}) \approx 2\,V_o\,e^{-i\,\pi/4}\,\mathfrak{Z}.$$

Thus the velocity for large $|\mathfrak{Z}|$ is

$$\overline{F'(\mathfrak{Z})} \approx 2 V_o\ e^{i\ \pi/4}$$

and $V = 2 V_o$. Now the complex potential (9) becomes

$$F(\mathfrak{Z}) = \frac{V}{2} \left[e^{-i\ \pi/4}(\mathfrak{Z} + \sqrt{\mathfrak{Z}^2 - 1}) + \frac{c^2\ e^{i\ \pi/4}}{\mathfrak{Z} + \sqrt{\mathfrak{Z}^2 - 1}} \right] .$$

Problems

6.1 Verify the mapping described in Figure 9.2 (iii).

6.2 A flat plate of width 2 is placed in a uniform flow with velocity $V\ e^{i\ \pi/3}$. Find the complex potential. Locate the stagnation points.

6.3 A uniform flow in the direction of the real axis with velocity V encounters the elliptical obstacle of Example 3. Find the complex potential describing the flow.

6.4 Find the complex potential describing the flow of a fluid which has uniform velocity V when it encounters a hurdle of height b as shown.

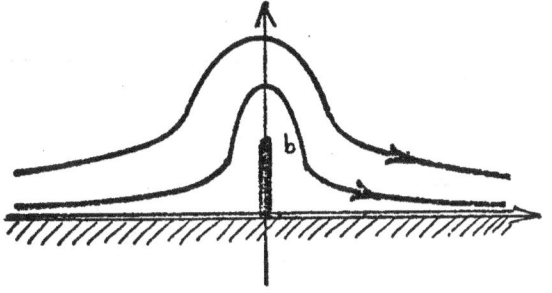

Figure 9.2   Joukowski transformation $\mathcal{z} = \frac{1}{2}\left(z + \frac{1}{z}\right)$

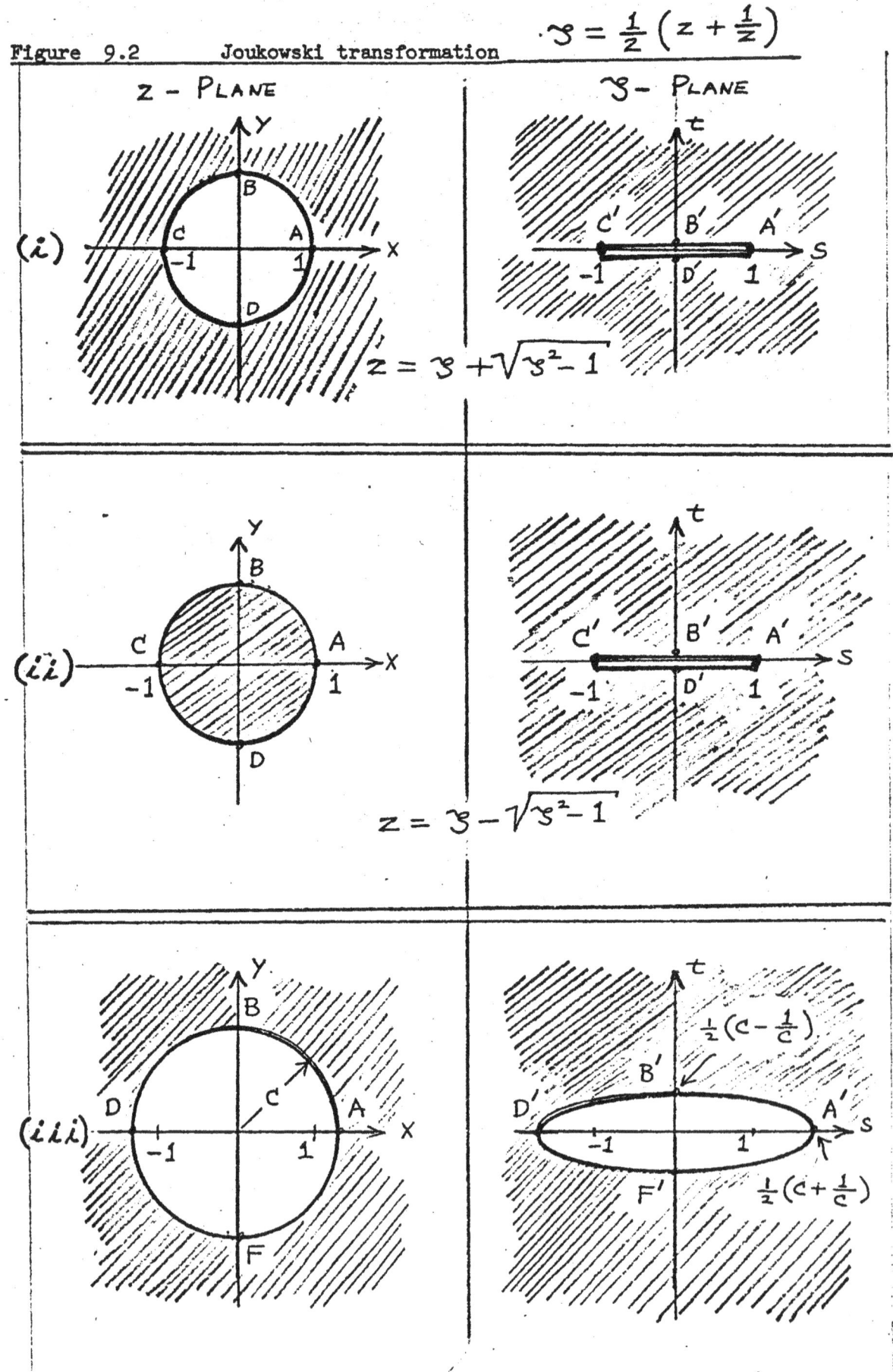

Figure 9.2 (continued)

z - PLANE   𝔖 - PLANE

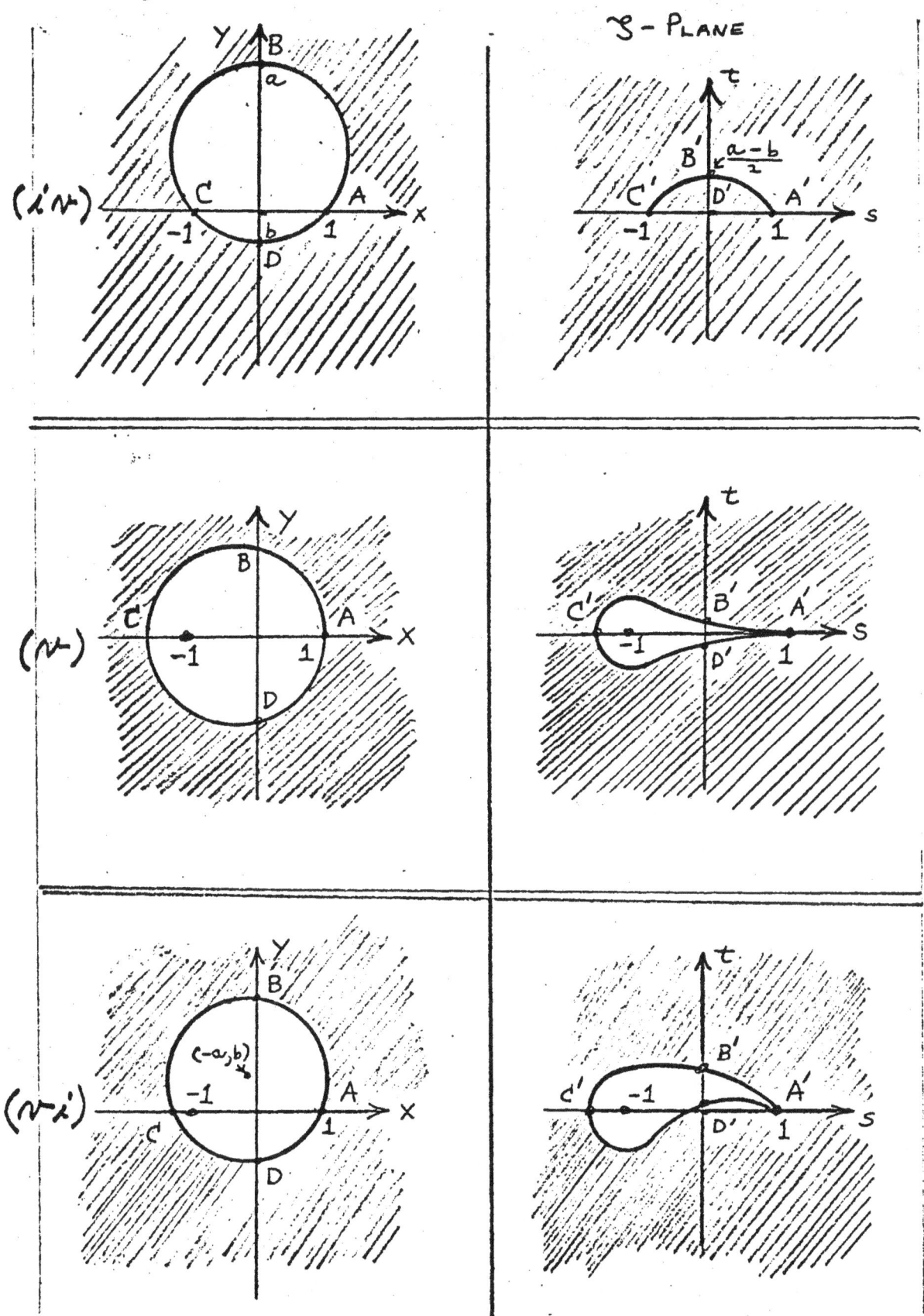

(iv)

(v)

(vi)

One application of the Joukowski transformation is to the study of air

flow past the wing of an airplane. Look at Figure 9.2 (vi) and observe that

the curve in the ζ- plane

resembles the cross-section

of an airplane wing. In

supplementary problems 9.6.3

and 9.6.4 the flow past these "Joukowski profiles" is briefly examined.

9.7 The electrostatic field

We have seen in the previous sections that an analytic function $f(z) =$

u (x, y) + i v(x, y) can be thought of as a convenient "package" for the

description of certain fluid flow problems. The analytic function also

provides us with a description of the two dimensional electrostatic field.

Suppose we have several parallel cylindrical metalic conductors in empty

space as shown. We assume that the cylinders are very long, so that

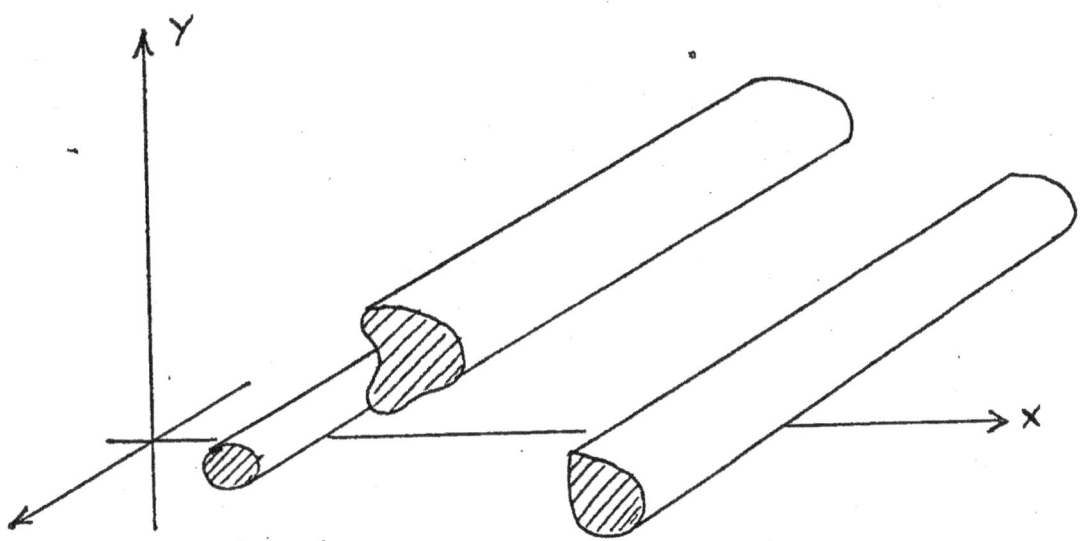

the electric field established is identical in any plane perpendicular to

these cylinders. Suppose electric charges are placed on each conductor.

An electrostatic potential u(x, y) is then established in the space

surrounding the conductors which has the properties:

(1) $\nabla^2 u(x, y) = 0$ in free space, $(u_{xx} + u_{yy} = 0)$

(2) u is a constant on the surface of each conductor.

Our knowledge of analytic functions can help us to select appropriate potential functions for specific problems. We illustrate this in the following examples.

Example 1

The positive and negative halves of the x-axis each represent infinite conducting plates separated from each other by insulation at the origin. The plate $x > 0, y = 0$ is kept at potential u = 3, while the plate $x < 0, y = 0$ is kept at potential u = -2. Find u in the upper half plane.

Solution

We seek a harmonic function in the upper half plane $y > 0$ that is 3 on the positive x-axis and -2 on the negative x-axis. The function $\theta = \arg(z)$ is the imaginary part of the function $\log z$ and is thus harmonic. Furthermore, θ takes on the value 0 on the positive real axis and π on the negative real axis. Can we find constants A and B such that

(1) $u = A\theta + B$

is the desired potential? When $\theta = 0$ and $u = 3$ we have from (1) that $B = 3$.

When $\theta = \pi$ and $u = -2$, (1) now reads $-2 = \pi A + 3$. Thus $A = -5/\pi$. The

desired potential is then

$$(2) \quad u = -\frac{5}{\pi} \theta + 3.$$

Since $\theta = \tan^{-1} y/x$, we have

$$(3) \quad u = -\frac{5}{\pi} \tan^{-1} y/x + 3, \quad 0 \le \tan^{-1} y/x \le \pi.$$

The electrostatic potential (2) is the real part of the function

$$f(z) = \frac{5}{\pi} i \log z + 3.$$

We call $f(z)$ the "complex electrostatic potential" or simply the "complex

potential."

Example 2

Three conductors on the x-axis, separated by insulation at $x = \pm 1$

are charged as shown

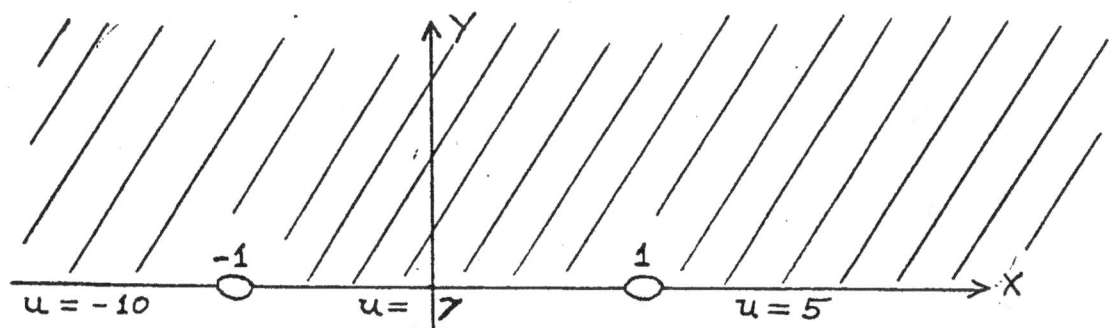

Determine u in the upper half plane and the complex potential.

<u>Solution</u>

We assume a solution of the form

(4) $u = A\theta_1 + B\theta_2 + C$

where A, B and C are constants to be determined and θ_1 and θ_2 are the angles

shown.

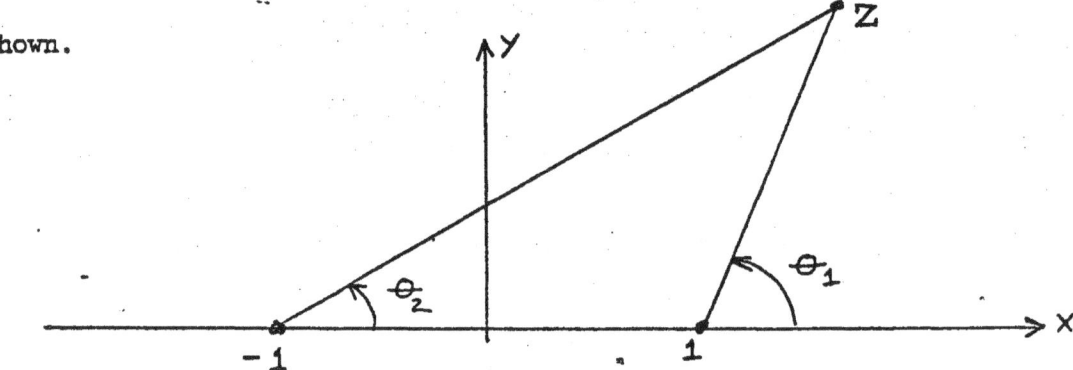

We know that u is a harmonic function because θ_1 is the imaginary part of

$\log(z - 1)$ and θ_2 is the imaginary part of $\log(z + 1)$.

From the given boundary conditions we see that u = 5 when

$\theta_1 = \theta_2 = 0$. Thus (4) becomes $5 = 0 + 0 + C$ and we see that C = 5.

When $\theta_1 = \pi$ and $\theta_2 = 0$ we see that u = 7. Thus (4) becomes $7 = \pi A + 0 + 5$

and we see that $A = 2/\pi$. Finally, when $\theta_1 = \theta_2 = \pi$, we have u = -10.

Now (4) becomes $-10 = \dfrac{2}{\pi} \pi + \pi B + 5$ or $B = -17/\pi$. Now we have

(5) $u = \dfrac{2}{\pi}\theta_1 - \dfrac{17}{\pi}\theta_2 + 5.$

Since $\theta_1 = \tan^{-1}\dfrac{y}{x-1}$ and $\theta_2 = \tan^{-1}\dfrac{y}{x+1}$ we have

$u = \dfrac{2}{\pi}\tan^{-1}\dfrac{y}{x-1} - \dfrac{17}{\pi}\tan^{-1}\dfrac{y}{x+1} + 5,$ where both angles are in the range $[0, \pi]$. The expression (5) is the real part of

(6) $f(z) = -\frac{2i}{\pi} \ln (z - 1) + \frac{17\,i}{\pi} \ln (z + 1) + 5$

and thus (6) is the desired complex potential.

Problems

7.1 Find the potential u(x, y) and the complex potential f(z) = u + i v in

the upper half plane when conductors separated by insulation on the x-axis

are charged to the potentials shown.

(a) $u = 2$ $u = 3$ → X
        0

(b) $u = 4$ $u = 0$ → X
      0   2

(c) $u = 2$ $u = 0$ $u = 2$ → X
       0   5

7.2 The positive real axis and the positive

imaginary axis are conducting plates separated

by insulation at the origin and are charged

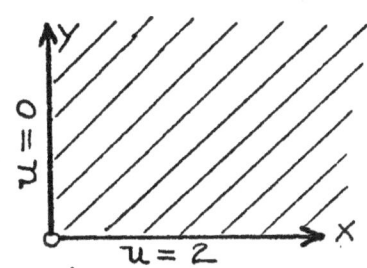

as shown. Determine the potential u(x, y) in the first quadrant. Also

find the complex potential.

● 7.3 Determine the potential in the

angular region shown where V_0 and

V_1 are constants. Also find the

complex potential.

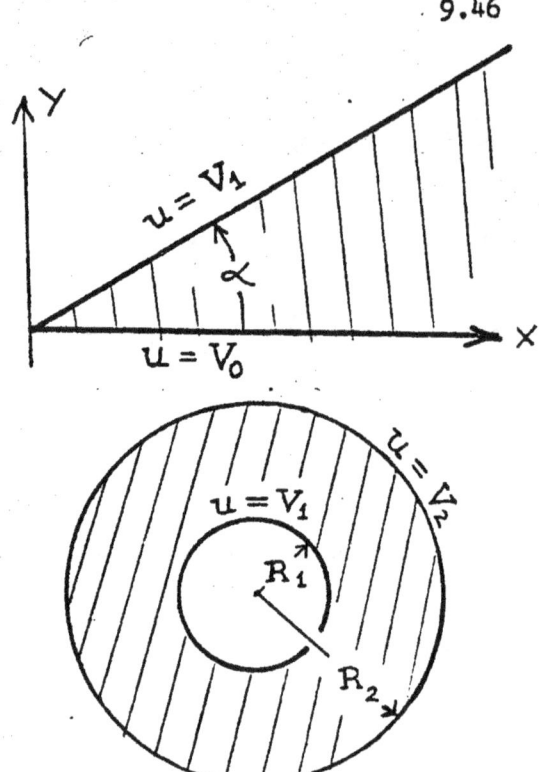

● 7.4 Determine the potential between

the two charged coaxial cylinders

shown.

We saw previously that $f(z) = m \log(z-z_0)$ described

a source of fluid flow. Here we imagined an infinitely long line piercing

the complex z-plane orthogonally at the point z_0 with fluid rushing out

radially in all directions. The quantity of fluid emerging from each

linear foot of the line source was $2\pi m$ units of volume per second.

Corresponding to these line sources of fluid flow we have line sources of

the electrostatic potential. Picture a straight wire conductor perpendicular

to the z -plane at z=0 on which an electric charge of density q units

of charge per linear unit of distance on the wire is evenly distributed.

The electrostatic potential generated by this line charge is

$$u = -2q \log r$$

and the complex potential is

(7) $f(z) = 2q \log z.$

We will not derive (7), since our principal interest is in the applications

of conformal mapping.

Corresponding to the velocity vector in fluid flow, we have the

electric field intensity vector $\vec{E}$. In general, a line charge q per unit

length would be subjected to the force $q\,\vec{E}$ per unit length. The electric

field intensity vector is described by the equations

$$\vec{E} = -\ \nabla u, \qquad \text{and}$$

(8) $\qquad \vec{E} = -\ \overline{f'(z)}.$

Example 3

A line of charge q_1 per unit length is at z = 0 and another of

charge q_2 per unit length is at z = 1. Find the potential, the complex

potential and the electric field intensity vector.

Solution

From (7) we can write down the complex potential at once as

(9) $\quad f(z) = -2q_1 \log z - 2 q_2 \log (z - 1)$ which is the sum of the

potentials due to the sources at $z = 0$ and at $z = 1$. Since the potential

u is given by $u = \text{Re } f(z)$ we have

$$u = -2q_1 \ln |z| - 2q_2 \ln |z - 1|$$

$$u = -2q_1 \ln \sqrt{x^2 + y^2} - 2q_2 \ln \sqrt{(x - 1)^2 + y^2}$$

$$u = -q_1 \ln (x^2 + y^2) - q_2 \ln ((x - 1)^2 + y^2).$$

The electric field intensity vector is

$$\vec{E} = - \overline{f'(z)}$$

$$\vec{E} = \frac{2q_1}{\bar{z}} + \frac{2q_2}{\bar{z} - 1}.$$

We have seen that the $\text{Re } f(z) = u$ describes the electrostatic

potential. What does $\text{Im } f(z) = v$ describe? In our study of fluid flow

the lines $v = c$ (c constant) were called stream lines and the velocity

vectors were tangent to these lines. In the study of the electrostatic

field the lines $\text{Im } f(z) = v = c$ are called <u>lines of force</u>. The vector

$\vec{E}$ is tangent to these lines.

<u>Example 4</u>

Suppose that in Example 3, $q_1 = q_2 = q$. Find the equations of the lines of force and sketch these lines and the equipotential lines.

Solution

Since $q_1 = q_2 = q$, the complex potential (9) becomes

$$f(z) = -2q \log z - 2q \log(z - 1)$$

$$f(z) = -2q \log z \, (z - 1)$$

$$f(z) = -2q \log \left| z \, (z - 1) \right| - 2q \arg (z(z - 1)) \, i.$$

But $z(z - 1) = x^2 - y^2 - x + i \, (2xy - y)$ and therefore the angle made by

$z(z - 1)$ is $\arg (z(z - 1)) = \tan^{-1} \dfrac{2xy - y}{x^2 - y^2 - x}$.

The imaginary part of the complex potential is given by

$$v(x, y) = -2q \tan^{-1} \frac{2xy - y}{x^2 - y^2 - x}$$

and thus the lines of force are obtained from $-2q \tan^{-1} \dfrac{2xy - y}{x^2 - y^2 - x} = c'.$

We get

$$\frac{2xy - y}{x^2 - y^2 - x} = \tan\left(-\frac{c'}{2q}\right) = C, \quad \text{where C is an arbitrary constant.}$$

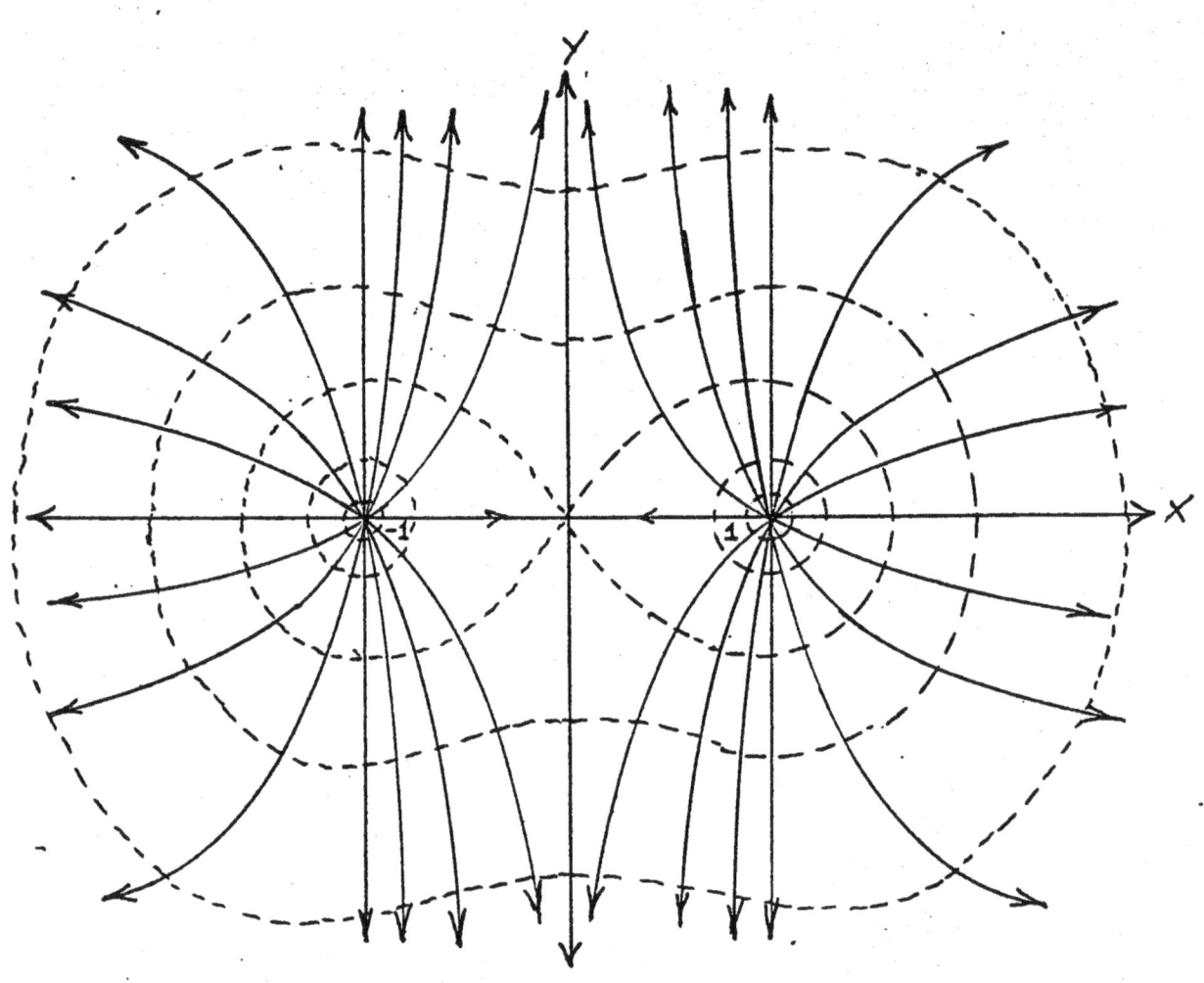

u(x,y) = constant ⎯ ⎯ ⎯ ⎯ ⎯ ⎯
equipotential lines

v(x,y) = constant ⎯⎯⎯⎯⎯⎯⎯⎯⎯
lines of force

Example 5

      Identical line charges q per unit length are placed at the points

$z = \pi n$, $(n = 0, \pm 1, \pm 2, \cdots)$.

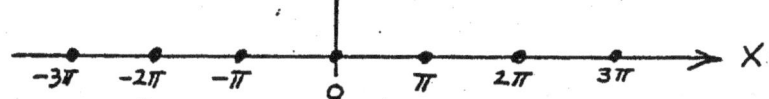

Find the complex potential at all points in the z-plane.

Solution

      This problem is the electrostatic analog of the fluid flow problem

considered in Example 5 of section 9.4. We need only replace the strength

m of the sources of fluid flow by $-2q$ to get our complex potential

$f(z) = -2q \log (\sin z)$.

Example 6

      A line of charge q per unit length is located at the point $z = i$

while the x-axis describes a grounded conductor (has potential $u = 0$).

Find (a) the complex potential, (b) the electrostatic potential, (c) the

lines of force, and (d) the electric intensity vector.

Solution

      The expression $-2q \log |z - i|$ describes the required line charge at $z = i$, but will not make the potential zero on the real axis. Suppose

we place a line of charge $-q$ at $z = -i$. Then the potential on the real

axis will be zero. Thus the required electrostatic potential is

$$u(x, y) = -2q \log |z - i| + 2q \log |z + i|$$

and the complex potential is

$$f(z) = -2q \log (z - i) + 2q \log (z + i)$$

$$f(z) = 2q \log \frac{z + i}{z - i} \; .$$

The charge at $-i$ is called the _image_ of the charge at i and the technique

used here is called the _method of images_.

Since

$$\frac{z + i}{z - i} = \frac{x + i(y + 1)}{x + i(y - 1)} = \frac{x^2 + y^2 - 1 + i\,2x}{x^2 + (y - 1)^2}$$

we have

$$f(z) = 2q \log \frac{|z+i|}{|z-i|} + 2qi \arg \left(\frac{z+i}{z-i} \right)$$

$$= 2q \log \frac{\sqrt{x^2 + (y+1)^2}}{\sqrt{x^2 + (y-1)^2}} + 2qi \tan^{-1} \left[\frac{2x}{x^2 + y^2 - 1} \right]$$

$$f(z) = q \log \left[\frac{x^2 + (y + 1)^2}{x^2 + (y - 1)^2}\right] + 2 q i \tan^{-1} \left[\frac{2x}{x^2 + y^2 - 1}\right]$$

which is the complex potential. We see at once that the electrostatic

potential is

$$u(x, y) = \text{Re } f(z) = q \log \left[\frac{x^2 + (y + 1)^2}{x^2 + (y - 1)^2}\right].$$

The lines of force are $v(x, y) = 2q \tan^{-1} \left[\frac{2x}{x^2 + y^2 - 1}\right] = c''$

or $\quad \frac{2x}{x^2 + y^2 - 1} = c'$.

These can be expressed as $\quad x^2 + y^2 - 1 = 2C x \quad (C = \frac{1}{C'})$

which reduce to

$$(x - C)^2 + y^2 = C^2 + 1$$

upon completing the square. These are circles with centers $(C, 0)$ on the

real axis passing through the line charges at $\pm$ 1.

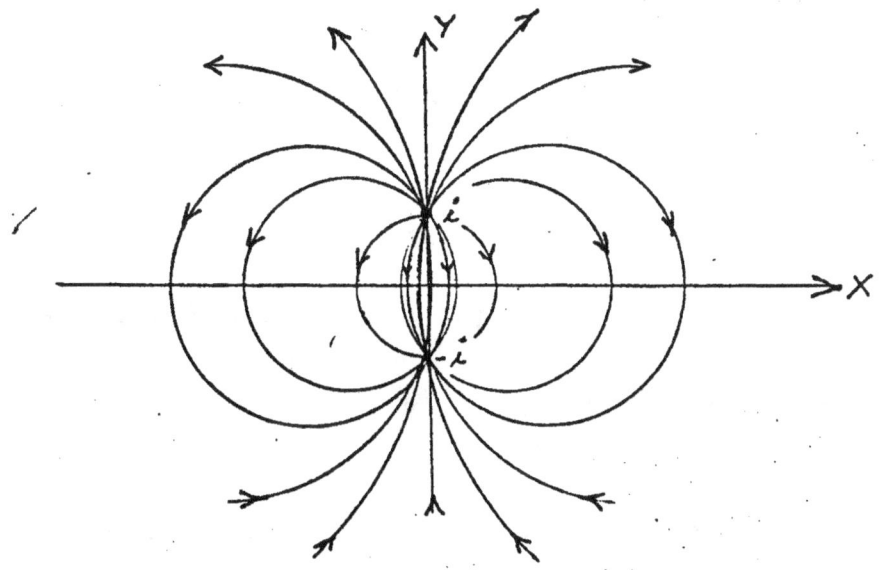

The electric intensity vector $\vec{E}$ is

$$\vec{E} = -\overline{f'(z)}$$

$$= \overline{\left(-\frac{1}{4}\left[\frac{2q}{z-i} + \frac{2q}{z+i}\right]\right)}$$

$$= \overline{\left(\frac{4\,qi}{z^2+1}\right)}$$

$$= \frac{-4q\,i}{\bar{z}^2+1}$$

Problems

In problems 24 through 29 find (a) the complex potential, (b) the electrostatic potential, (c) the lines of force and (d) the electric intensity vector for each of the given distributions of line charges.

7.5 A line of charge q per unit length is at z = 1 and a line of charge -q per unit length is at z = -1.

7.6 Identical lines of charge q per unit length are located at z = $\pi/2 + \pi n$, where n is any integer.

7.7 Identical lines of charge q per unit length are located at z = i n, where n is any integer.

7.8 Identical lines of charge -q per unit length are located at z = πn, where n is any integer.

● 7.9   Lines of charge q per unit length are located at z = n, while lines

of charge -q per unit length are located at z = ½ + n where n is any integer.

● 7.10   A line of charge q per unit length is located at z = 1 while the

imaginary axis is a grounded conductor.

● 7.11   A line of charge q per unit

length is located at (a, b) in the

first quadrant. The positive real

and imaginary axes are grounded

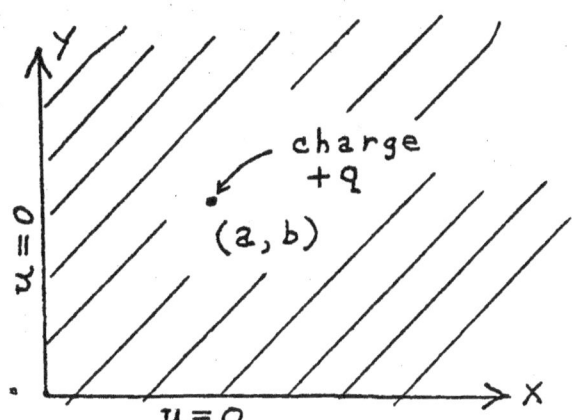

conductors. Find the complex potential and the electrostatic potential.

A useful transformation for mapping flow patterns and electrostatic

potentials is obtained through the function $z = e^{\mathcal{Y}}$. This function maps the

upper half of the z-plane onto the infinite channel $0 \leq \text{Im}(\mathcal{Y}) \leq \pi$ on

the $\mathcal{Y}$-plane. It is useful to visualize this mapping in the following way.

Visualization of the Mapping $z = e^{\zeta}$

(1) Imagine the upper half of

the z-plane as a rubber

sheet.

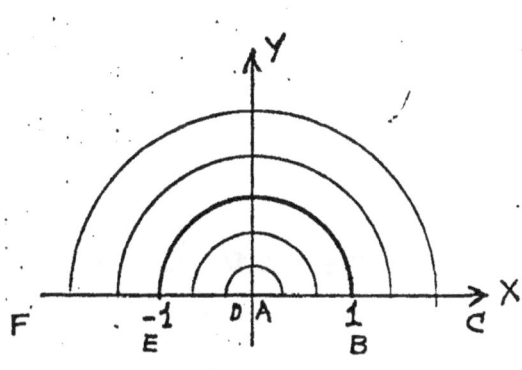

(ii) Break the rubber sheet at the origin.

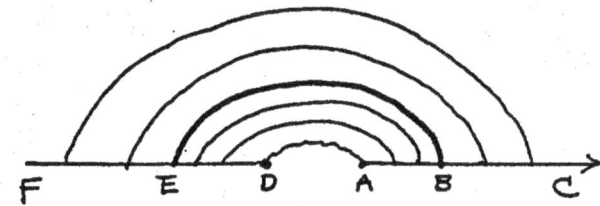

(iii) Hold the edge ABC fixed and rotate the edge DEF until it is parallel to ABC and the strip is of width π.

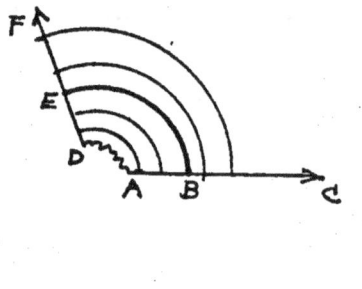

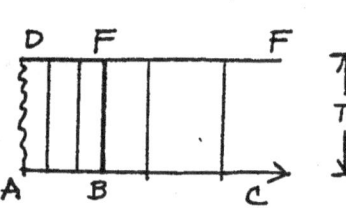

(iv) Grab both the ends at A and D and stretch them to the left to infinity. The line segment BE falls on the imaginary $\mathfrak{J}$ axis.

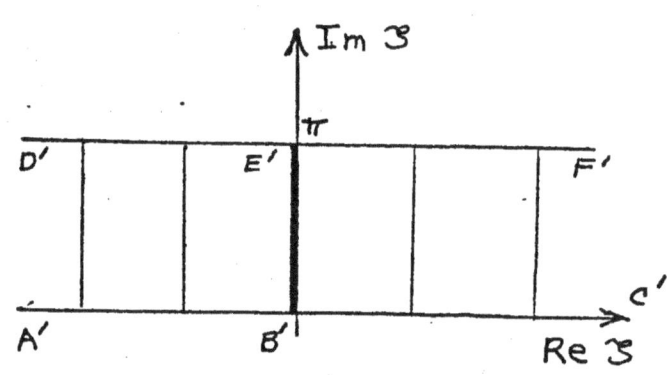

To see that this is the correct mapping the reader should demonstrate that the real axis on the $\mathfrak{J}$-plane ($\mathfrak{J} = s$), $t = 0$, $-\infty < s < \infty$) maps onto the positive real axis of the z-plane and that the line $\mathfrak{J} = s + i\pi$ ($-\infty < s < \infty$) maps onto the negative real axis of the z-plane under the mapping $z = e^{\mathfrak{J}}$. Also the reader should show that the line segment $\mathfrak{J} = i t$ ($0 \leq t \leq \pi$) maps onto the upper half of the unit circle on the z-plane.

Example 7

The diagram shows an infinite channel of width π bound on both

sides by grounded conductors. At the

point $\mathcal{S} = i$ $\pi/2$ there is a line of

charge q. Determine the potential

function.

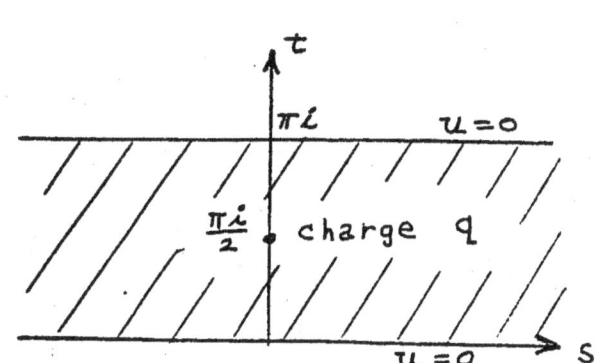

Solution

The mapping function $z = e^{\mathcal{S}}$ transforms the channel on the

$\mathcal{S}$-plane into the upper half of the z-plane. The line source at

$\mathcal{S} = i$ $\pi/2$ maps onto z = i. The lower boundary of the channel maps onto

the positive x-axis and the upper boundary onto the negative x-axis. If

we solve the problem of having a line source of charge q at z = i while the

x-axis is at zero potential, the function $z = e^{\mathcal{S}}$ will map that solution

onto the desired channel in the $\mathcal{S}$-plane. In Example 6 we saw that the

solution of this problem in the z-plane is the complex potential.

$$f(z) = 2q \log \frac{z + i}{z - i} .$$

Thus the solution to the present problem is obtained by setting $z = e^{\zeta}$

and obtaining the complex potential

(10) $f(z) = 2q \log \dfrac{e^{\zeta} + i}{e^{\zeta} - i}$.

To find the potential $u(s, t)$ and the family of lines of force $v(s, t) = C$

we must find the real and imaginary parts of $(e^{\zeta} + i)/(e^{\zeta} - i)$.

$$\frac{e^{\zeta} + i}{e^{\zeta} - i} = \frac{e^{s} e^{it} + i}{e^{s} e^{it} - i} \cdot \frac{e^{s} e^{-it} + i}{e^{s} e^{-it} + i}$$

$$= \frac{e^{2s} - 1 + i\, e^{s} (e^{it} + e^{-it})}{e^{2s} + 1 + i\, e^{s} (e^{it} - e^{-it})}$$

$$= \frac{e^{2s} - 1 + i\, 2 e^{s} \cos t}{e^{2s} + 1 - 2 e^{s} \sin t}$$

Now we see that

$$\left| \frac{e^{\zeta} + i}{e^{\zeta} - i} \right| = \frac{\sqrt{(e^{2s} - 1)^2 + 4e^{2s} \cos^2 t}}{e^{2s} + 1 - 2e^{s} \sin t}$$

and

$$\arg \left(-\frac{e^{\zeta} + i}{e^{\zeta} - i} \right) = \tan^{-1} \frac{2e^{s} \cos t}{e^{2s} - 1} .$$

Combining these last two expressions with (10) gives us the electrostatic

potential

$$u(s, t) = q \log \left[\frac{(e^{2s} - 1)^2 + 4e^{2s} \cos^2 t}{\left(e^{2s} + 1 - 2e^{s} \sin t \right)^2} \right]$$

and the family of lines of force

$$v(s, t) = 2q \tan^{-1} \frac{2e^s \cos t}{e^{2s} - 1} = C'.$$

This last expression reduces to

$$\frac{e^s \cos t}{e^{2s} - 1} = C$$

where C is an arbitrary constant.

Problems

● 7.12 .   Move the line of charge q to the point $\Im = bi$ in Example 7 and

find the resulting electrostatic potential.

● 7.13   The diagram shows a line of

charge q   at $\Im = bi$ in a channel of

width "a" bound by two grounded con-

ducting plates.   Find the potential.

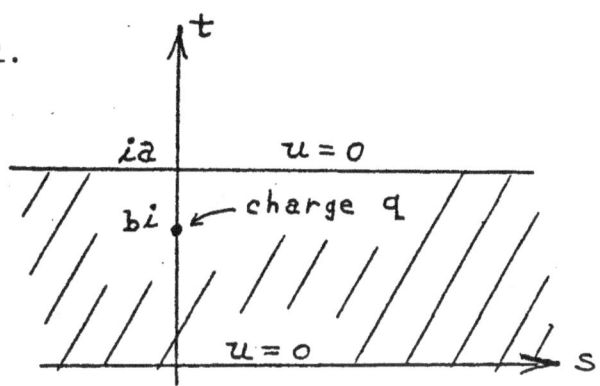

● 7.14   The diagram shows a line of

charge q between two infinite

grounded conducting plates making

angle $\pi/4$. Find the complex potential.

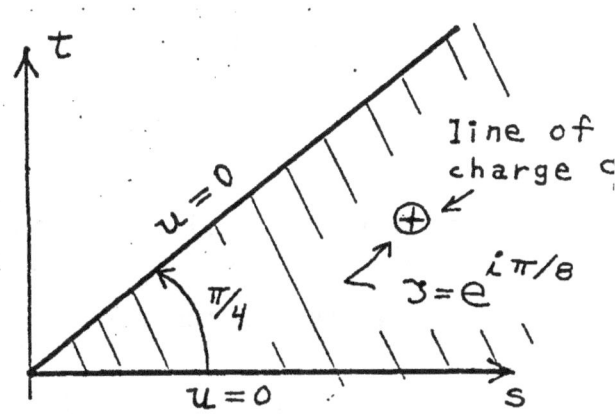

9.8 Heat flow

We have stud^{ied} problems in fluid flow and the electrostatic potential.

We now study two dimensional, steady state temperature distributions. The

figure shows the cross sectional

boundary of a cylindrical piece of

material that extends to infinity

in a direction perpendicular to

the z-plane. Every plane section

of the cylinder parallel to the

z-plane looks like this. Suppose

the temperature distribution is

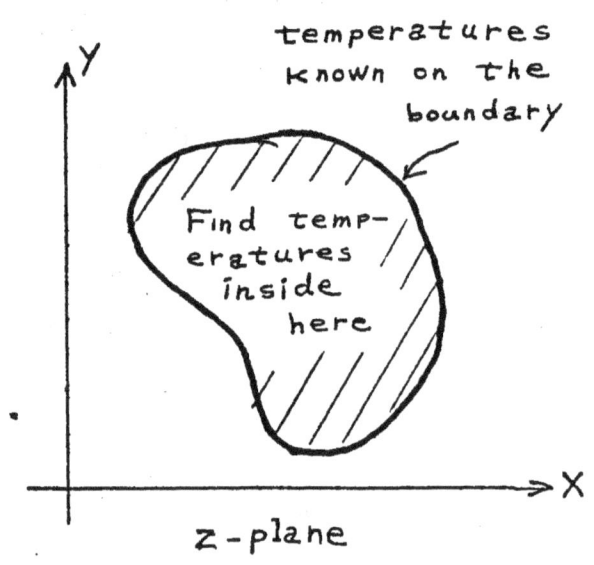

known on the boundary, and that the temperatures are not changing with

time. What is the temperature at each point inside the cylinder? Let

$u(x, y)$ denote the temperature at the point (x, y). In section 9.2

we saw that $u(x, y)$ satisfies Laplace's equation and is thus the real part

of some analytic function $f(x) = u(x, y) + i\, v(x, y)$. We call $f(z)$ the

complex temperature. The problem at hand is to select the appropriate

function $f(z)$ whose real part $u(x, y)$ matches the given temperatures on the boundary of the cylinder.

Example 1

Consider the vertical slab of
material shown. The left face of the
slab is maintained at the constant
temperature 30° and the right face is
at 70° (centigrade). Find the temperature distribution throughout the
slab.

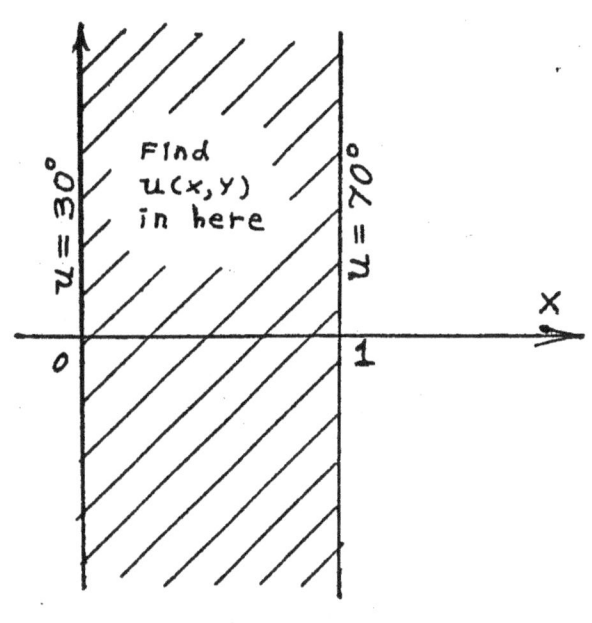

Solution

We must find a harmonic function u(x, y) such that

$$u(0, y) = 30, \text{ and } u(1, y) = 70,$$

Of course, we also require that u(x, y) be "nice" (have no singularities
or discontinuities) inside the slab $(0 \leq x \leq 1)$. It is clear that the
function u(x, y) should be independent of the variable y, since the
boundary conditions do not change with y. Suppose we try a simple function
of the form u = Ax + B, where both A and B are constants. This is a

harmonic function since it is the real part of the analytic function

$f(z) = Az + B$.   The boundary condition on the left edge of the slab tells

us that when x = 0, u = 30 and thus $30 = A \cdot 0 + B$

The boundary condition on the right edge of the slab reads u = 70   when

x = 1 yielding $70 = A \cdot 1 + B$.

The simultaneous solution of these two equations is B = 30, and A = 40.

We have then the required temperature distribution

$$u(x, y) = 40\ x + 30$$

and the complex temperature

$$f(z) = 40\ z + 30.$$

We notice at once that this problem and its solution is exactly

equivalent to the following problem from electrostatics: "Find the potential

u(x, y) inside the channel bound by conducting plates at x = 0 and at x = 1

when the left plate is maintained at potential 30 and the right plate is

maintained at potential 70." We have only a change in terminology, but no

change in the required mathematical analysis.

If $f(z) = u(x, y) + i\ v(x, y)$ is the complex temperature, then

the level lines of the temperature Re $f(z) = u(x, y) = C$ (constant)   are called "isotherms", and the level lines of $\text{Im } f(z) = v(x, y) = C$ are

called "lines of flux". Lines of flux are lines along which heat flows.

In the previous example the isotherms are the vertical lines $x = c$ while

the lines of flux are the horizontal lines $y = c$. Notice that heat flows

along the flux lines from right to left (since the right edge has the

higher temperature).

Table 9.1 shows the similarity in the applications considered in this chapter.

Remark on Example 1

A mathematical problem is called "well posed" if it has a solution,

and only that one solution. We say a solution exists and is unique when

the problem is well posed. Is the problem in Example 1 well posed? Clearly

a solution exists, for we found one: $u = 40 x + 30$. But is it unique? Un-

fortunately, no. Consider the funciton

$$\sin \pi z = \sin \pi x \cosh \pi y + i \cos \pi x \sinh \pi y.$$

The real part of this function is harmonic, and zero for $x = 0$ and for $x = 1$.

Thus the function

$$U(x, y) = 40 x + 30 + \sin \pi x \cosh \pi y$$

TABLE 9.1   COMPARISON OF TOPICS IN THE TWO DIMENSIONAL, STEADY STATE STUDY OF FLUID FLOW, ELECTROSTATICS AND HEAT FLOW

FLUID FLOW	ELECTROSTATICS	HEAT FLOW
velocity potential $u(x, y)$	electrostatic potential $u(x, y)$	temperature $u(x, y)$
complex potential $f(z) = u(x, y) + i\,v(x, y)$	complex electrostatic potential $f(z) = u(x, y) + i\,v(x, y)$	complex temperature $f(z) = u(x, y) + i\,v(x, y)$
stream lines $v(x, y) = c$	lines of force $v(x, y) = c$	lines of heat flux $v(x, y) = c$
equipotential lines $u(x, y) = c$	equipotential lines $u(x, y) = c$	isotherms $u(x, y) = c$
velocity vector $\vec{V} = \vec{\nabla}u = \overline{f'(z)}$	electric intensity $\vec{E} = -\vec{\nabla}u = -\overline{f'(z)}$	heat flux $\vec{H} = -K\vec{\nabla}u = -K\,\overline{f'(z)}$ $K =$ conductivity
at the boundary of an obstacle $v(x, y) = c$ and $\dfrac{\partial u(x, y)}{\partial n} = 0$ (n normal to the obstacle's surface)	At the boundary of a conductor $u(x, y) = c$	along the boundary of insulation material $\dfrac{\partial u(x, y)}{\partial n} = 0$ (n normal to the surface of the insulation)

is also a solution to the problem. Why does the solution to this problem

fail to be unique? The reason for this is our failure to specify what is

happening to the temperature across the top and the bottom of the slab.

Since our slab goes to infinity in the y direction, we note that there

is really no top and bottom to the slab. Nevertheless, we can think of

the slab as having finite, though very large, height. Notice that because

$$\lim_{y \to \pm \infty} \cosh y = \infty$$

the solution $U(x, y)$ approaches infinity for large y. This means there

is a great heat source at the top and bottom of the slab. If we had re-

quired that the solution $u(x, y)$ be <u>bounded</u> throughout the slab, the

solution $U(x, y)$ would have been eliminated and the solution of the problem

would then be unique! For this reason, we will add the restriction that

the temperature be bounded when the material under question extends to

infinity.

Besides having the temperature specified on the boundary of the

material whose internal temperature is to be determined, we can also specify

that a portion of the boundary is <u>insulated</u>. The presence of insulation means that heat cannot flow into,

or out of that portion of

the boundary. Thus an insulated

segment of the boundary coincides

with a line of flux $v(x, y) = C$.

This also means that $\dfrac{\partial u(x, y)}{\partial n} =$

0, where n is normal to the in-

sulated surface. We will identify an insulated segment of a boundary by

the symbol 〰〰〰 .

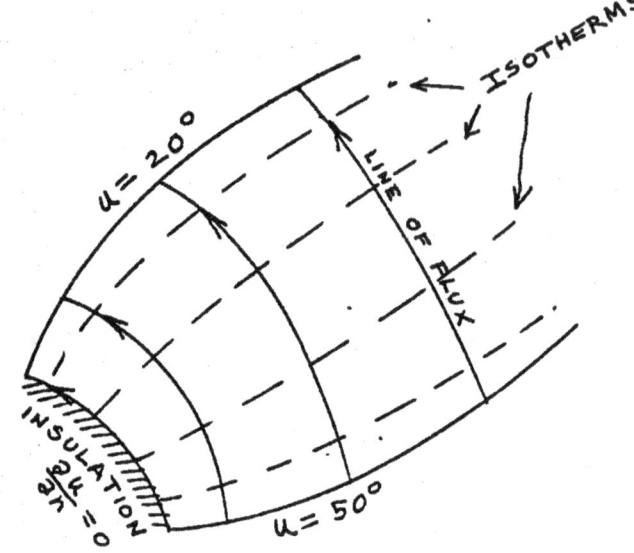

Example 2

Find the bounded temperatures

inside the slab shown.

Solution

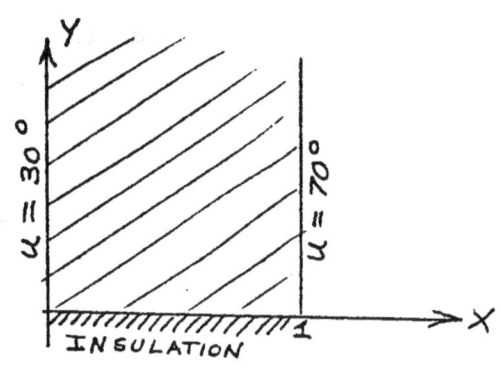

Our solution $u = 40x + 30$ to Example 1 gives the required

temperatures on the left and the right boundaries. Also, the insulated

segment of the x-axis is a line of flux $y = 0$. Thus $u = 40x + 30$ is

the required solution. Note also that along the insulation, the normal derivative $\dfrac{\partial u}{\partial y}$ is zero.

Problems

● 8.1 Find the bounded temperature

distribution inside the slab shown.

Also find the complex temperature,

the lines of flux and the isotherms.

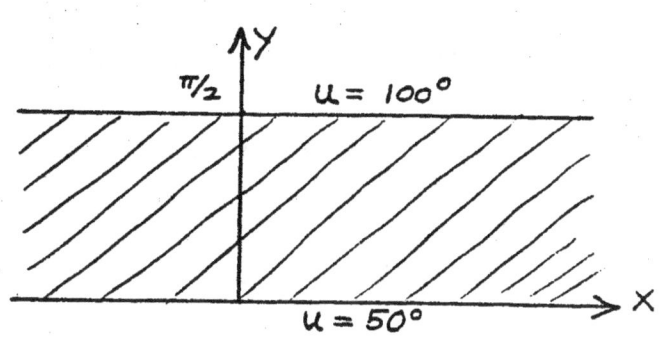

● 8.2 Find the bounded temperature

inside the slab shown.

● 8.3 Find the bounded temperature

inside the region shown.

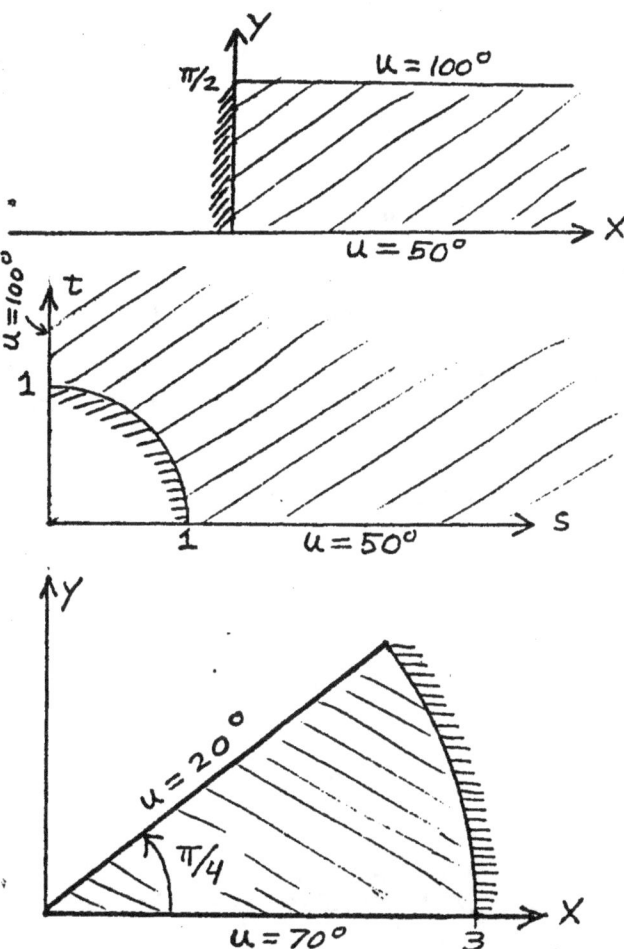

● 8.4 Find the temperature inside

the region shown. Also find the

complex potential and the lines

of flux.

Another useful transformation is the function

(1)

$z = \sin \mathfrak{Z}$

$z = \sin s \cosh t + i \cos s \sinh t$.

It is convenient to visualize this mapping in the following way:

(1) Consider the semi-infinite

strip shown and imagine it is

made of rubber in the $\Im$-plane

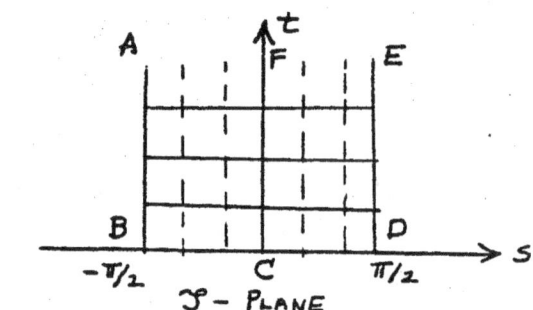

(ii) Fold the lines AB and

ED down as shown.

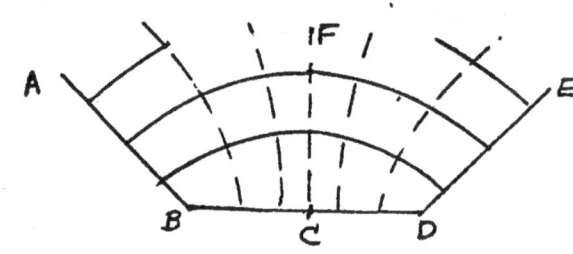

(iii) When these lines reach

the x-axis, the formerly hor-

izontal lines t = c should be

ellipses and the vertical lines

should be hyperboles. s = C.

Both these families of conics

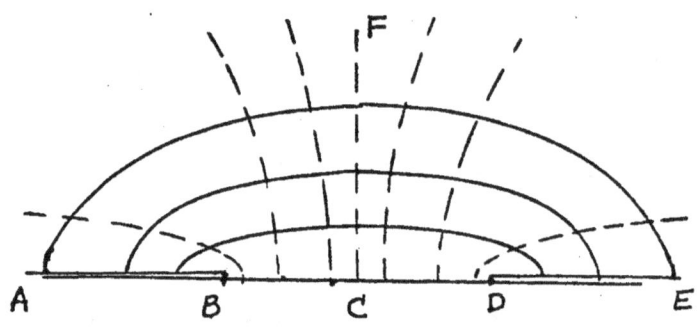

have the points B and D as foci and are thus called confocal ellipses

and hyperbolas.

(iv) Finally, move the points

B and D to -1 and +1 re-

spectively on the z-plane.

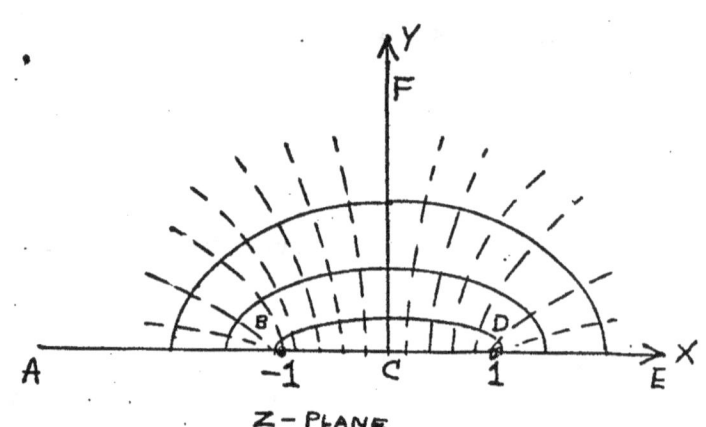

The following relations derived from $z = \sin \mathcal{J}$ are useful:

Transformations from (s, t) to (x, y):

(2) $x = \sin s \cosh t$

(3) $y = \cos s \sinh t$.

Transformations from (x, y) to (s, t):

(4) $s = \sin^{-1} \left[\tfrac{1}{2} \left(\sqrt{(x + 1)^2 + y^2} - \sqrt{(x - 1)^2 + y^2} \right) \right]$

$$(\text{here } -\pi/2 \leq \sin^{-1} [\ ] \leq \pi/2)$$

(5) $t = \cosh^{-1} \left[\tfrac{1}{2} \left(\sqrt{(x + 1)^2 + y^2} + \sqrt{(x - 1)^2 + y^2} \right) \right]$.

The equation of the confocal ellipses is

(6) $\dfrac{x^2}{\cosh^2 t} + \dfrac{y^2}{\sinh^2 t} = 1$. The equation of the confocal hyperbolas is

(7) $\dfrac{x^2}{\sin^2 s} + \dfrac{y^2}{\cos^2 s} = 1$.

A method of deriving these relations is outlined in supplementary problem 9.8.9.

Example 3

Find the bounded temperatures

inside the shaded region shown

with the indicated boundary

conditions.

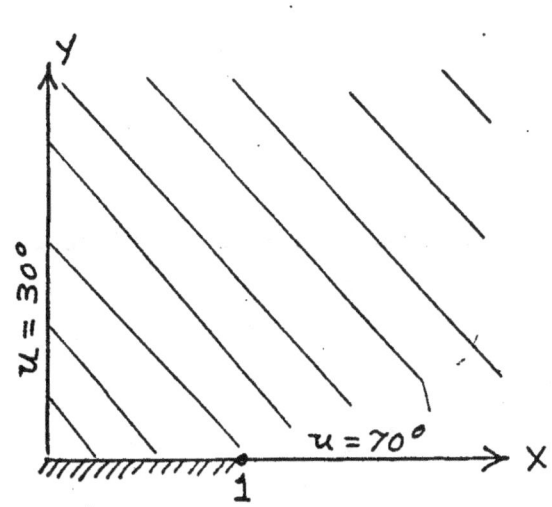

Solution

Return to Example 2. Notice that similar boundary conditions occured

in that problem. Thus we need only employ the necessary transformations

to map the semi-infinite slab considered in Example 2 onto the first

quadrant to get the solution of the present problem.

The following two mappings are used:

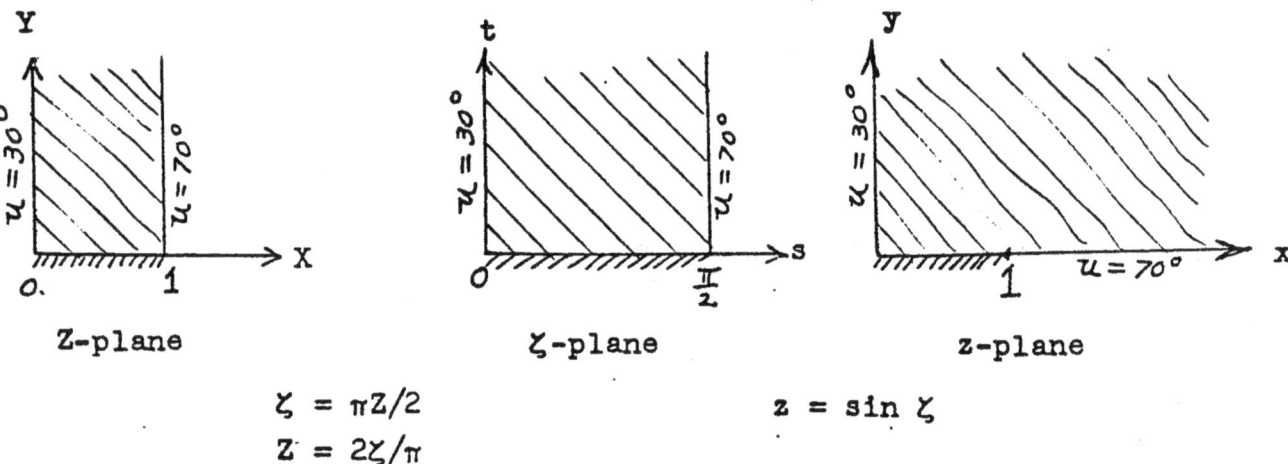

$$\zeta = \pi Z/2$$
$$Z = 2\zeta/\pi$$

$$z = \sin \zeta$$

The solution to the problem in the Z-plane is u = 40 X + 30 as we saw in

Example 2. (Here we have replaced x by X.) The first transformation is

the simple magnification $\zeta = \pi Z/2$ or $Z = 2\zeta/\pi$. This means that

$$X + i Y = \frac{2s}{\pi} + i \frac{2t}{\pi}$$

and consequently

$$X = \frac{2s}{\pi}.$$

The solution for u now becomes

$$u = \frac{80\,s}{\pi} + 30.$$

The second mapping $z = \sin \zeta$ requires the use of relation (4) to transform

this temperature into

$$u = \frac{80}{\pi} \sin^{-1}\left[\tfrac{1}{2}\left(\sqrt{(x+1)^2 + y^2} - \sqrt{(x-1)^2 + y^2}\right)\right] + 30$$

which is the desired solution.

Problems

Find the bounded temperature distribution $u(x, y)$ inside each of the

shaded regions having the indicated boundary conditions.

● 8.5

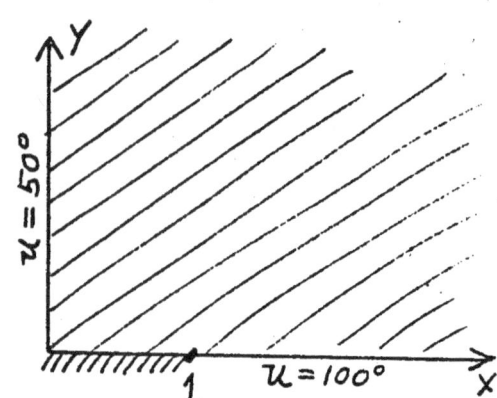

● 8.6

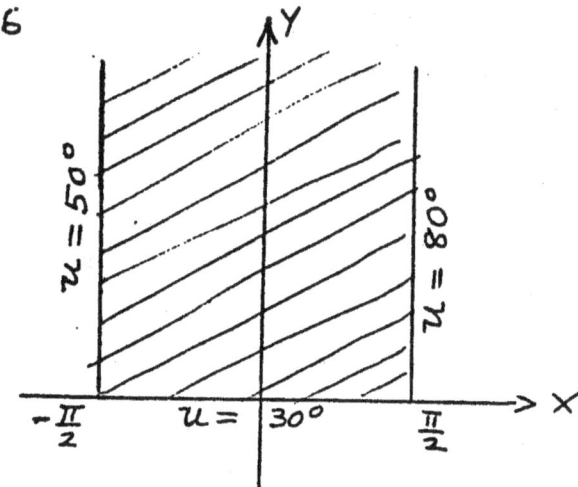

● 8.7

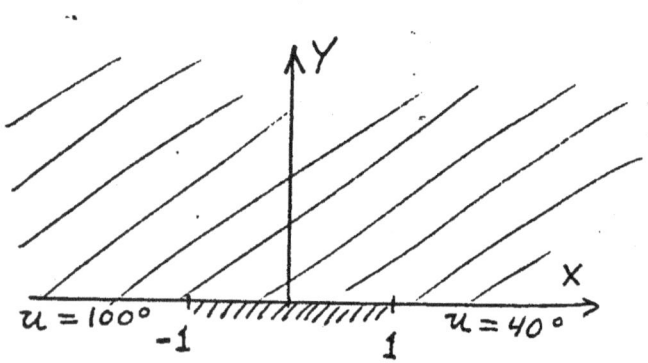

● 8.8

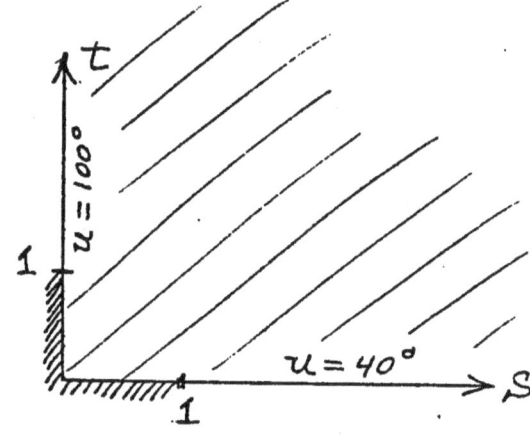

9.9 The bilinear transformation

The mapping function

(1) $w = \dfrac{az + b}{cz + d}$, where $ad - bc \neq 0$,

is called a bilinear transformation, a linear fractional transformation and a Möbius transformation. It is a simple, yet important mapping and we shall study its effects in this section.

The special case of (1) in which $c = 0$ reduces to the linear transformation

(2) $w = az + b$

which we examined in section 5.5. Recall that the mapping produced by (2) can be thought of as a magnification (by the factor $|a|$), followed by a rotation (through the angle arg (a)), followed by a translation (by the vector b).

Another special case of (1) is the so called "inversion transformation"

(3) $w = \dfrac{1}{z}$.

The inversion defines a conformal mapping at all points except, of course, $z = 0$. If we write $z = re^{i\theta}$

then we have $w = \dfrac{1}{z} = \dfrac{1}{r} e^{-i\theta}$.

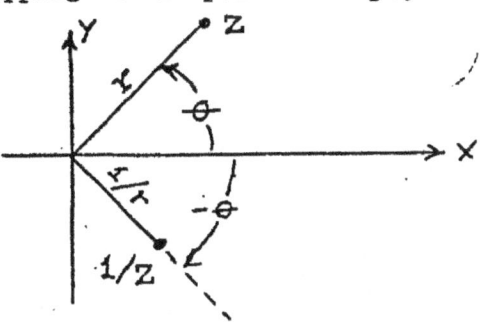

This last relation tells us that if we know the polar coordinates of a complex number, for example $z = 2 + 2i = 2\sqrt{2}e^{i\,\pi/4}$ has coordinates $r = 2\sqrt{2}$ and $\theta = \pi/4$, then the coordinates of $1/z$ are easily found:

radius: $\dfrac{1}{r} = \dfrac{1}{2\sqrt{2}}$, and

angle: $-\theta = -\dfrac{\pi}{4}$,

An important property of the inversion transformation (3) is that it preserves circles. This means that a circle in the z-plane is always transformed into a circle in the w-plane and vice versa. Here, however, we must also allow circles of infinite radius, which are straight lines. A method for proving this property is outlined in SP9.9.12. The following example illustrates a geometric technique for mapping a circle under the inversion transformation.

Example 1

Find the mapping of the circle shown under $w = \dfrac{1}{z}$.

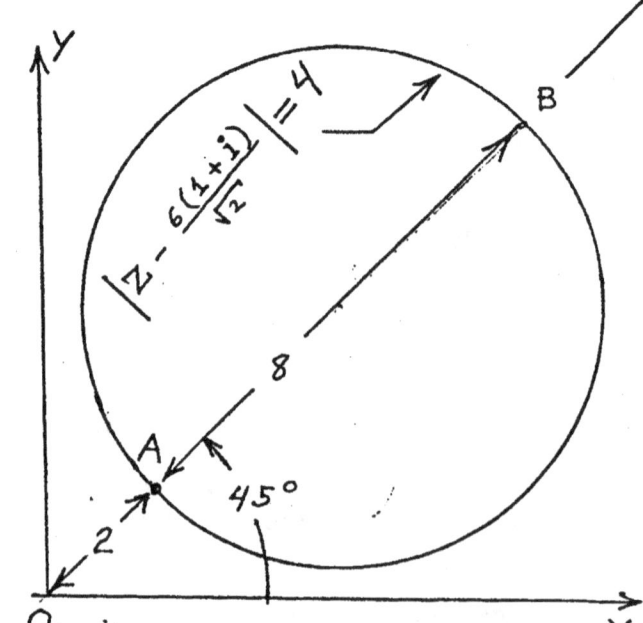

Solution

First note that the line O A B D making angle 45° maps onto the line

D′B′A′O′ shown making

angle $- 45^\circ$. The

lengths $\overline{D'A'}$ and

$\overline{D'B'}$ are 1/3 and

1/10 respectively.

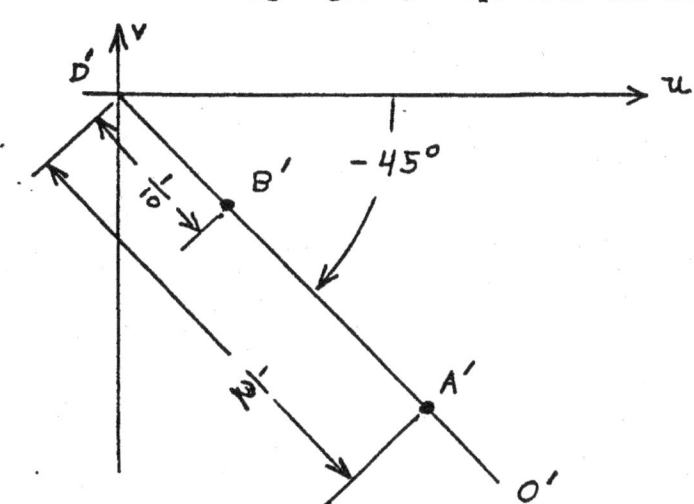

Since the mapping 1/z is conformal, and since the circle intersects the

line segment $\overline{AB}$ at right angles in the z-plane then it must intersect

$\overline{A'B'}$ at right angles in the w-plane.

Thus $\overline{A'B'}$ is a diameter of

our circle and we draw it

as shown.

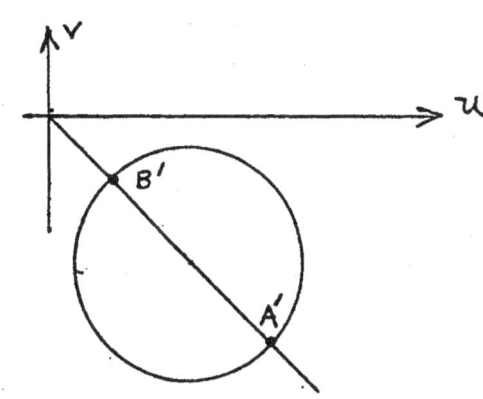

The diameter of the circle in the w-plane is $1/2 - 1/10 = 2/5$ and thus

the radius is 1/5. The center of this circle is at distance $1/10 + 1/5 =$

3/10 from the origin. The center is thus at the point

$$\frac{3}{10} e^{-i \pi/4} = \frac{3(1 - i)}{10 \sqrt{2}} ,$$

and the equation of the circle is

$$\left| w - \frac{3(1-i)}{10\sqrt{2}} \right| = \frac{1}{5} \ .$$

Example 2

Find the mapping of the circle

$$\left| z - 1 - \sqrt{3}\,i \right| = 4 \quad \text{under}$$

$$w = 1/z.$$

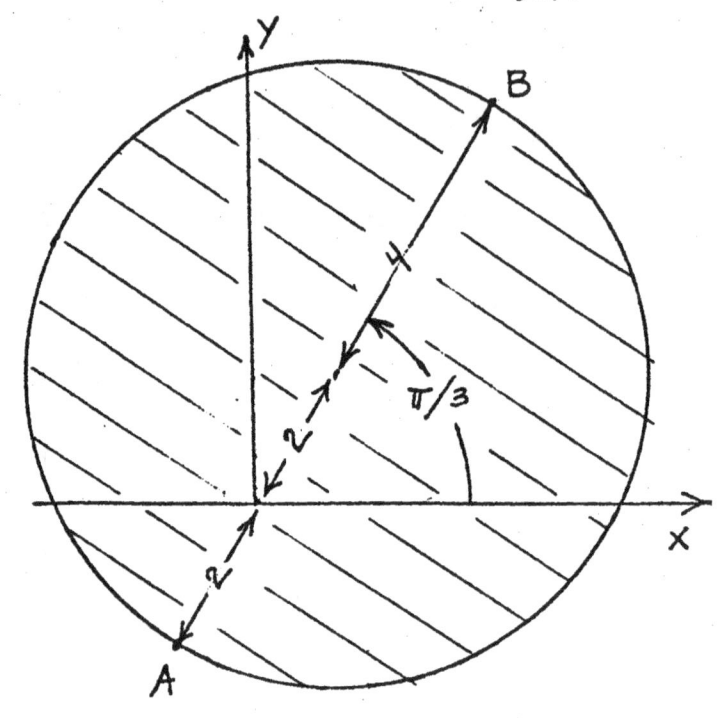

Solution

The center of the circle in the z-plane is

$$1 + \sqrt{3}\,i = 2e^{i\,\pi/3}$$

and the radius is 4. The points A and B

$$A: \quad 2e^{4\pi i/3}$$

$$B: \quad 6e^{\pi i/3}$$

are on a diameter passing through the origin of the z-plane and these map

onto the points

$$A': \quad 1/2\ e^{-4\pi i/3}$$

$$B': \quad 1/6\ e^{-\pi i/3}$$

which are shown in the w-plane.

Since A'B' is a diameter of the

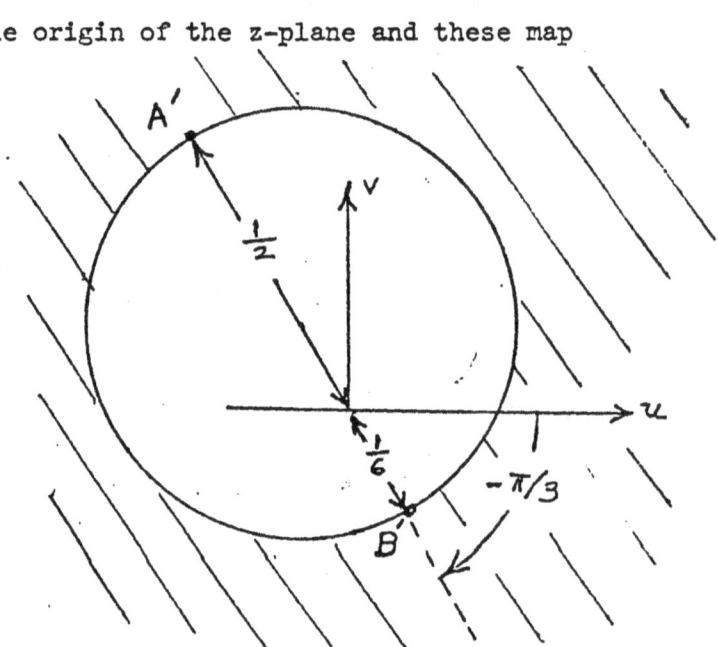

new circle, we can draw the image as shown. Notice that the <u>interior</u> of

the circle in the z-plane maps onto the <u>exterior</u> of its image circle in

the w-plane. How do we know this? We can use a test point. Notice that

z = 0 is in the shaded region in the z-plane, and therefore its image

w = 1/0 = ∞ must be in the shaded portion of the w-plane.

To get the equation of our image circle in the w-plane we note that

its diameter is

$$1/2 + 1/6 = 2/3$$

and thus its radius is 1/3. The center must be at a distance

$$1/2 - 1/3 = 1/6$$

from the origin and is thus at the point

$$1/6 \, e^{i2\pi/3} = 1/12 \, (-1 + \sqrt{3}\, i) \ .$$

The desired equation is then

$$\left| w - 1/12(-1 + \sqrt{3}\, i) \right| = 1/3 \ .$$

Problem

9.1   Find the mapping of each of the following regions under the inversion

transformation w = 1/z.

(a) $|z - 2| \leq 1$,   (b) $|z - 4i| \leq 2$,

(c) $|z - 1| \leq 1$,   (d) $|z + 1 - i\sqrt{3}| \leq 2$,

(e) $|z + 1 + i| \leq 3\sqrt{2}$,   (f) $\text{Re}(z) \leq 2$.

We can now study the mapping properties of the general bilinear function,

$$w = \frac{az + b}{cz + d} , \quad ad - bc \neq 0.$$

Multiplying by $cz + d$ gives

$$czw - az + dw - b = 0,$$

and solving for z yields

$$(4) \quad z = \frac{-dw + b}{cw - a} .$$

Relation (4) is the inverse of (1) and we see that it too is a bilinear

transformation. Together, the relations (1) and (4) show that the bilinear

transformation is a one to one mapping of the extended z-plane onto the ex-

tended w-plane. Notice that

$$z = - d/c \longleftrightarrow w = \infty$$
$$w = a/c \longleftrightarrow z = \infty$$

The derivative of (1) is

$$(5) \quad \frac{dw}{dz} = \frac{ad - bc}{(cz + d)^2} .$$

If $ad - bc = 0$, then (5) is zero for all z and w must be a constant function. This would be a very uninteresting transformation,

$$w = c \text{ (constant)} ,$$

and thus we shall always assume that

(6) $ad - bc \neq 0.$

With (6), (5) shows that $\frac{dw}{dz}$ is never zero and thus (1) defines a con-

formal mapping for all z except $z = -d/c$ which maps into $w = \infty$ anyway.

In previous problems and examples we mapped circular regions under

the inversion transformation $w = 1/z$. We now show how to map regions with

the general bilinear transformation

$$w = \frac{az + b}{cz + d} .$$

Factor away $\frac{a}{c}$ to get

$$w = \frac{a}{c} \cdot \frac{z + b/a}{z + d/c} .$$

Next add and subtract d/c in the numerator

$$w = \frac{a}{c} \cdot \frac{z + d/c + b/a - d/c}{z + d/c}$$

and simplify to get

$$w = \frac{a}{c} + \frac{bc - ad}{c^2} \cdot \frac{1}{z + d/c} .$$

Call $A = a/c$, $B = (bc - ad)/c^2$, and $d/c = D$ so that we have

(7) $w = A + B \cdot \dfrac{1}{z + D}$.

Consider now a region in the z-plane to be mapped to the w-plane by the bilinear transformation (1). Using (7) we can reduce this involved mapping to a sequence of simple mappings as follows:

(i) Translate the region by the vector D.

(ii) Apply the inversion transformation.

(iii) Magnify the region by the factor $|B|$.

(iv) Rotate the region through the angle arg (B).

(v) Finally, translate the region through the vector A.

Since each of the above five steps preserves circles, we see that the general bilinear transformation (1), like the inversion $w = 1/z$, preserves circles.

Example 3

Find the mapping of the unit circle $|z| \leq 1$ under the bilinear transformation.

(8) $w = -i \left(\dfrac{z - 1}{z + 1} \right)$.

Solution

We first convert w to the form $w = A + B \dfrac{1}{z + D}$

by the procedure specified above. We have

$$w = -i \left(\frac{z-1}{z+1} \right)$$

$$w = -i \; \frac{z + 1 -1 -1}{z + 1}$$

$$w = -i \left(1 - \frac{2}{z+1} \right)$$

(9) $w = -i + 2i \cdot \dfrac{1}{z+1}$.

We start with

the circle shown.

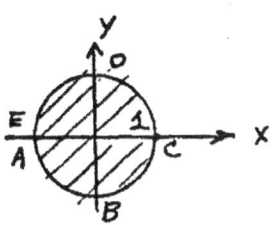

(i) The translation z + 1

yields

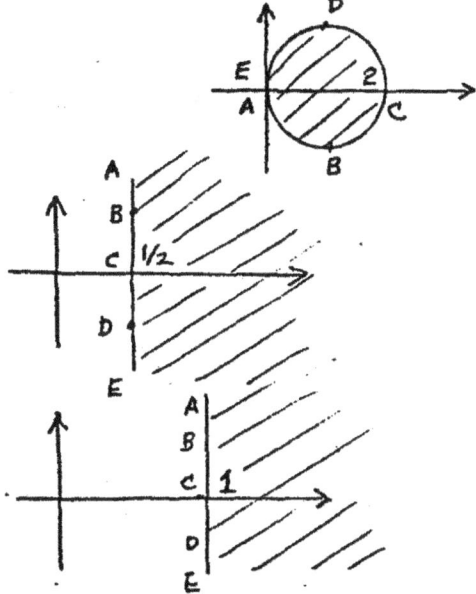

(ii) The inversion yields

(iii) The magnification by

$|2i| = 2$ gives

(iv) The rotation through

the angle

$\arg(2i) = \pi/2$ yields

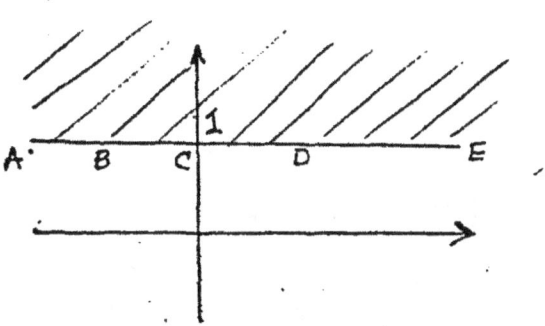

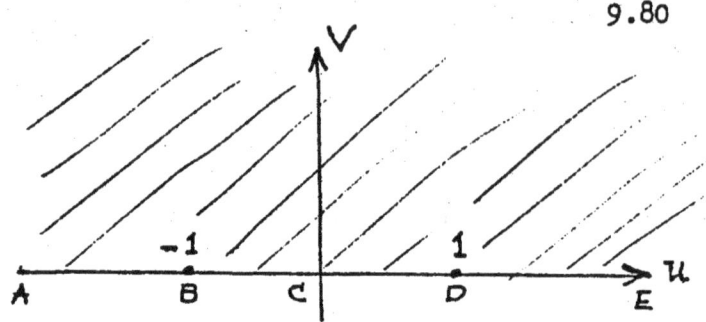

(v) Finally the translation

by the vector -i gives the

region shown.

We see that (8) maps the unit circle $|z| \leq 1$ onto the upper half plane

Im $w \geq 0$.

Problems

● 9.2   Map the regions (a) $|z| \leq 2$, and (b) $|z - 1| \leq 1$ under the trans-

formation $w = -i \left(\dfrac{z - 1}{z + 1} \right)$.

● 9.3   Map the region $\text{Im}(z) < 0$ under the transformation $w = \dfrac{z + 2}{z - 3}$.

A general mapping of the upper half plane $\text{Im}(z) \geq 0$ onto the unit circle

$|w| \leq 1$ is given by the transformation

(9) $$w = e^{i\alpha}\, \frac{z - a}{z - \bar{a}}$$

where "a" is a complex number in the upper half of the z-plane. To see

this, look at the modulus of both sides of (9).

$$|w| = \left| e^{i\alpha}\, \frac{z - a}{z - \bar{a}} \right|$$

$$|w| = \left| \frac{z - a}{z - \bar{a}} \right|$$

(10) $|w| = \dfrac{|z - a|}{|z - \bar{a}|}$

The diagram shows the numerator

and denominator of (10). Since

$|z - a| \leq |z - \bar{a}|$ when z is in

the upper half plane, and since

$|x - a| = |x - \bar{a}|$ when z = x

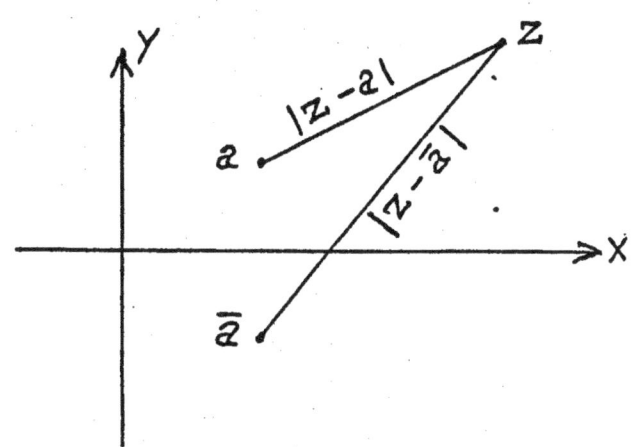

(is on the real axis), we see that Im(z) $\geq$ 0 is mapped onto $|w| \leq 1$.

We can use the bilinear transformation to solve a number of problems in

fluids, electrostatics and heat flow. The following examples illustrate

a few of these problems and their solutions.

Example 4

Find the temperature

distribution throughout

the region shown with

the given boundary

temperatures u = 20°

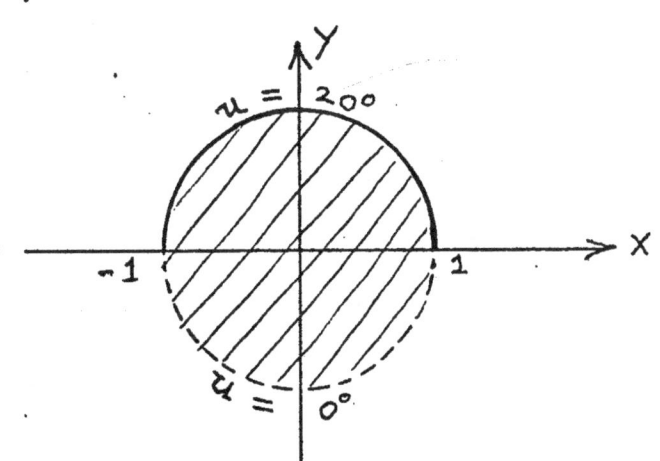

for $0 < \theta < \pi$ and u = 0° for $-\pi < \theta < 0$. Find the temperature at the

points z = 0, and z = i/2.

Solution

The mapping function

$$\mathcal{Y} = -i \left(\frac{z-1}{z+1} \right) \text{ maps}$$

the region $|z| \leq 1$ onto

the upper half of the

$\mathcal{Y}$- plane. (See Example 3.)

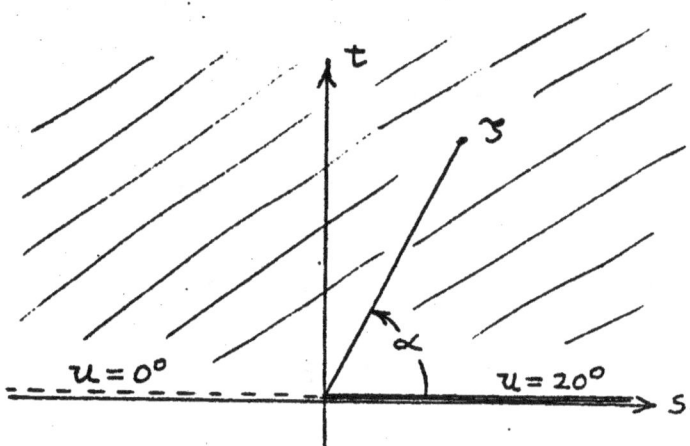

The upper semi-circular boundary at 20° in the z-plane maps onto the

positive s axis while the lower boundary at 0° maps onto the negative

s axis. The bounded temperature distribution in the $\mathcal{Y}$-plane can be

found by assuming that $u = A\alpha + B$, where $\alpha = \tan^{-1}(t/s)$ and $0 \leq \tan^{-1}(t/s) \leq \pi$.

$$) \leq \pi .$$

The boundary conditions generate the equations

$$20 = 0 \cdot A + B$$
$$0 = \pi \cdot A + B$$

which have the solution $B = 20$, $A = -20/\pi$ and thus

(11) $u = \dfrac{20}{\pi} \tan^{-1} \left(\dfrac{t}{s} \right) + 20.$

To find the solution in the circle on the z-plane we must write s and t

in terms of x and y. The appropriate transformations are found by writing

$$\mathcal{Y} = -i\left(\frac{z-1}{z+1}\right)$$

$$s + i\,t = -i\left(\frac{x-1+i\,y}{x+1+i\,y}\right)$$

$$s + i\,t = -i\left(\frac{x-1+i\,y}{x+1+i\,y}\right)\cdot\left(\frac{x+1-i\,y}{x+1-i\,y}\right)$$

$$s + i\,t = -i\left(\frac{x^2+y^2-1+2\,y\,i}{(x+1)^2+y^2}\right)$$

$$s + i\,t = \frac{2y+(1-x^2-y^2)\,i}{(x+1)^2+y^2}$$

Equating real and imaginary parts we get

$$(12)\quad s = \frac{2y}{(x+1)^2+y^2}\quad\text{and}\quad t = \frac{1-x^2-y^2}{(x+1)^2+y^2}$$

Substituting (12) into (11) we get

$$u(x,\,y) = -\frac{20}{\pi}\,\tan^{-1}\left(\frac{1-x^2-y^2}{2y}\right)+20,\qquad 0\le \tan(\ )\le \pi,$$

which *is* the required temperature distribution. The temperature at $(0,\,0)$ is

$$u = -\frac{20}{\pi}\,\tan^{-1}(\infty)+20$$

In the range from 0 to π, $\tan(\pi/2)=\infty$

and thus

$$u(0,\,0) = -\frac{20}{\pi}(\pi/2)+20 = 10^{\circ}$$

The temperature at $(0,\,\tfrac{1}{2})$ is

$$u(0,\,\tfrac{1}{2}) = -\frac{20}{\pi}\,\tan^{-1}(3/4)+20$$

$$= 15.903^{\circ}.$$

Example 5

A line of charge q per

unit length is located at

$z = \frac{1}{2}$ while the circle $|z| = $

1 is a grounded conductor.

Find the complex potential

inside $|z| \leq 1$ and the electrostatic potential.

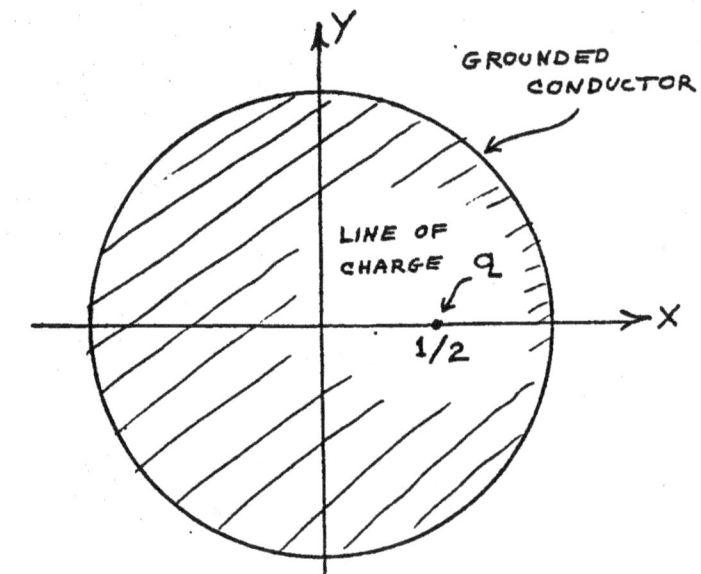

Solution

The mapping function

$$\mathcal{J} = -i \left(\frac{z - 1}{z + 1} \right)$$

described in Example 3 maps the conducting boundary onto the real $\mathcal{J}$-axis

and the line of charge at $z = \frac{1}{2}$ onto the point $\mathcal{J} = i/3$.

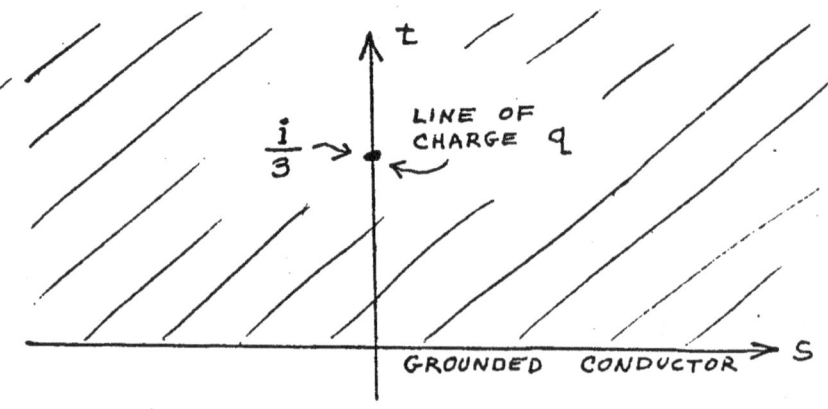

Placing an image line of charge -q per unit length at $\mathcal{I} = -\frac{1}{3}$ we get

$$f = -2q \log\left(\mathcal{I} - i/3\right) + 2q \log\left(\mathcal{I} + i/3\right)$$

$$f = 2q \log \frac{\mathcal{I} + i/3}{\mathcal{I} - i/3}$$

for the complex potential in the $\mathcal{I}$-plane. Using the bilinear transformation

we map this complex potential onto the unit circle $|z| = 1$ as

$$f(z) = 2q \log \frac{-i\left(\frac{z-1}{z+1}\right) + \frac{i}{3}}{-i\left(\frac{z-1}{z+1}\right) - \frac{i}{3}}$$

$$f(z) = 2q \log \frac{2z - 1}{z - 2}.$$

The electrostatic potential is $u = \mathrm{Re}\, f(z)$ and thus

$$u(x, y) = 2q \log \left|\frac{2z - 1}{z - 2}\right|$$

$$u(x, y) = q \log \frac{(2x - 1)^2 + 4y^2}{(x - 2)^2 + y^2}.$$

Example 6

Find the temperatures inside the

semi-circular region with the given

boundary values.

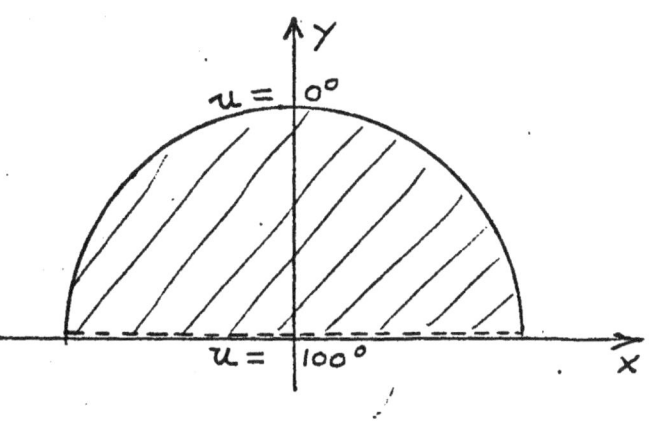

Z - PLANE

Solution

The mapping function $\mathcal{I} = -i \frac{z - 1}{z + 1}$

described in Example 3 maps this region

onto the first quadrant of the $\Im$-plane.

The lower boundary on the z-plane maps

onto the positive t-axis and the upper

boundary maps onto the positive s-axis.

The solution in the $\Im$-plane is

$$u = \frac{200}{\pi} \tan^{-1}\left(\frac{t}{s}\right).$$

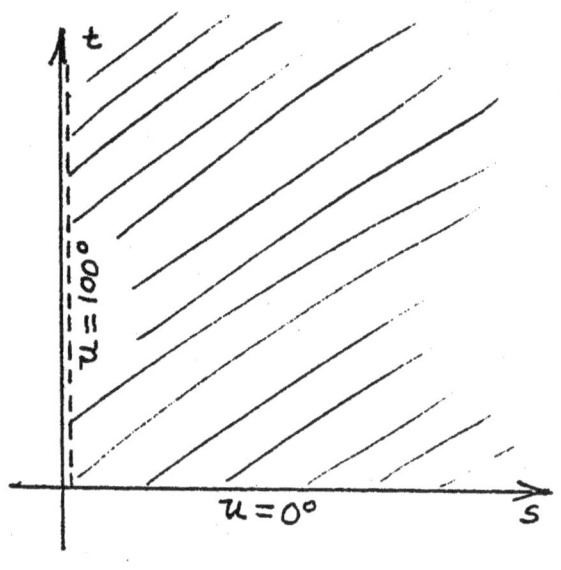

$\Im$ – PLANE

Using the relations (12) we have

$$u = \frac{200}{\pi} \tan^{-1} \frac{1 - x^2 - y^2}{2y}$$

where $0 \leq \tan^{-1}(\ \ ) \leq \pi/2$, for our solution.

Problems

Find the temperature distribution throughout each of the shaded regions

due to the temperatures maintained on the boundaries.

● 9.4

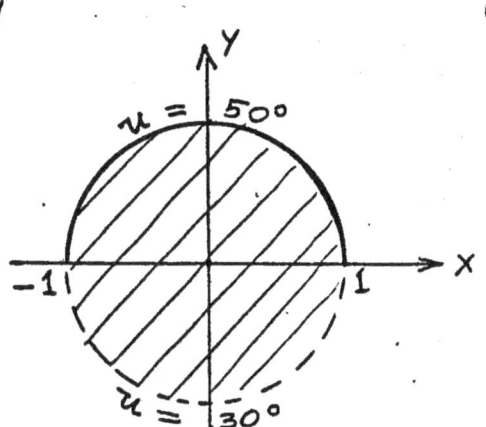

● 9.5

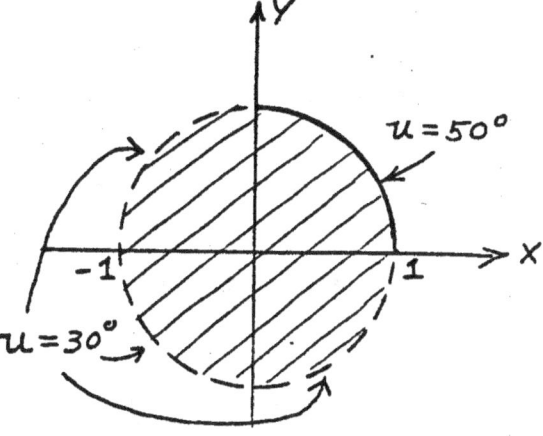

● 9.6

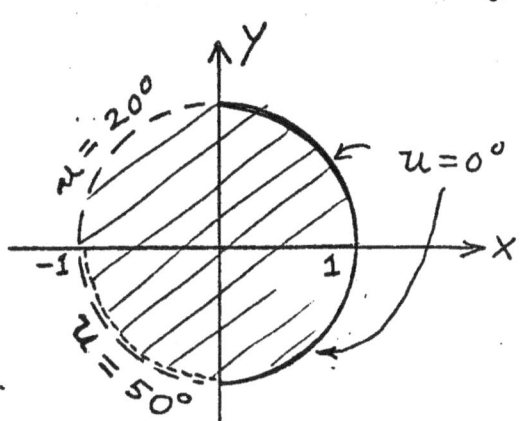

● 9.7

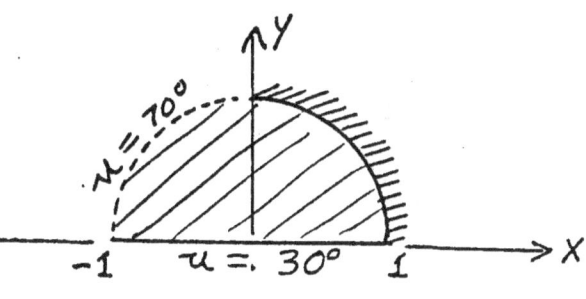

● 9.8

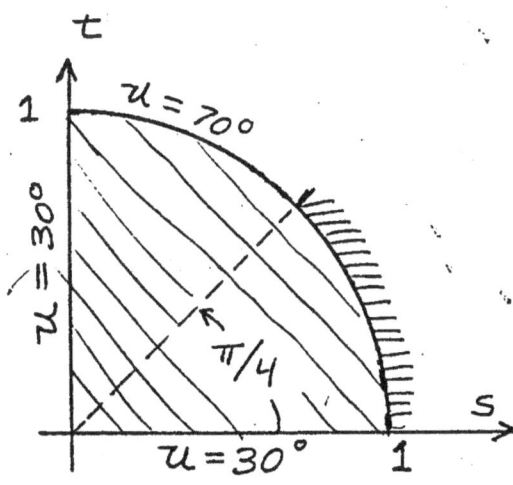

● 9.9   The boundary of the disc $|z| \leq 1$ represents a grounded conductor.

A line of charge q per unit length is placed at $z = 1/4$. Find the

complex potential that results.

9.10 More properties of the bilinear transformation

In this section we introduce additional properties of the bilinear

transformation

(1) $w = \dfrac{az + b}{cz + d}$, $ad - bc \neq 0.$

Property I   Mapping three given points into three given points.

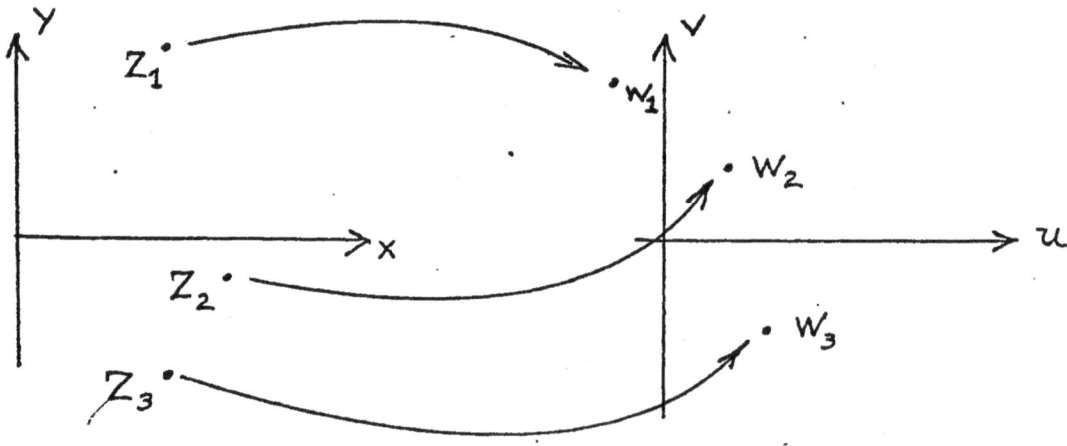

We can always find a bilinear function (1) which maps any three given

distinct points z_1, z_2, z_3 onto three specified distinct corresponding

points w_1, w_2, w_3.

Property II   Invariance of the cross ratio

If z_1, z_2, z_3, z_4 are distinct points in the z-plane which map onto the corresponding distinct points w_1, w_2, w_3, w_4 in the w-plane by the bilinear transformation (1), then

$$(2) \quad \frac{(z_4 - z_1)(z_2 - z_3)}{(z_2 - z_1)(z_4 - z_3)} = \frac{(w_4 - w_1)(w_2 - w_3)}{(w_2 - w_1)(w_4 - w_3)} .$$

The left side of (6) is called the cross ratio of z_1, z_2, z_3, z_4. If we wish to compute the bilinear transformation mentioned in Property I, we simply replace z_4 by z and w_4 by w in (2).

A method for deriving Property II is outlined in supplementary problem SP9.10.6.

Example 1

Find a bilinear transformation mapping the points 1, i, -1 in the z-plane into the points 0, 1, 2 in the w-plane, respectively.

Solution

We use the cross ratio (2) and set $z_1 = 1$, $z_2 = i$, $z_3 = -1$, $z_4 = z$, $w_1 = 0$, $w_2 = 1$, $w_3 = 2$ and $w_4 = w$.

We get
$$\frac{(z - 1)(i + 1)}{(i - 1)(z + 1)} = \frac{(w - 0)(1 - 2)}{(1 - 0)(w - 2)} .$$

Since $\dfrac{i + 1}{i - 1}$ $=$ $-\dfrac{(1 + i)}{(1 - i)}\dfrac{(1 + i)}{(1 + i)}$ $=$ $-\dfrac{2i}{2}$ $= -i,$

the above expression reduces to

$$-i\left(\frac{z - 1}{z + 1}\right) = -\frac{w}{w - 2}$$

Solving for w we get

$$i(w - 2)(z - 1) = w(z + 1)$$

which simplifies to

$$((1 - i)z + 1 + i) \; w = -2i(z - 1)$$

and finally

$$w = \frac{-2i(z - 1)}{((1 - i)z + 1 + i} \; .$$

Example 2

Find a bilinear transformation mapping the points $0, 1, \infty$ on the z-plane

into $\infty, i, 1$ on the w-plane, respectively.

Solution

Again we use (2) with $z_1 = 0$, $z_2 = 1$, $z_3 = \infty$, $z_4 = z$, $w_1 = \infty$, $w_2 = i$,

$w_3 = 1$ and $w_4 = w$. We get

$$\frac{(z - 0)(1 - \infty)}{(1 - 0)(z - \infty)} = \frac{(w - \infty)(i - 1)}{(i - \infty)(w - 1)} \; .$$

We interpret the terms

$$\frac{1 - \infty}{z - \infty} \qquad \text{and} \qquad \frac{w - \infty}{i - \infty}$$

simply as "1" because $\displaystyle\lim_{\zeta \to \infty} \frac{a - \zeta}{b - \zeta} = 1$.

We now have

$$z = \frac{i - 1}{w - 1}$$

and solving for w we get

$$w = \frac{z - 1 + i}{z}$$

Problem

● 10.1   Find the bilinear transformation mapping the three given points in

the z-plane onto the specified points in the w-plane.

(a)

z	w
-2	i
0	1
2	-i

(b)

z	w
0	∞
1	1
-1	0

(c)

z	w
0	∞
∞	0
1	1

Definition of symmetric points

Two points P_1 and P_2 are said

to be symmetric with respect to

the circle C having center at 0

if and only if

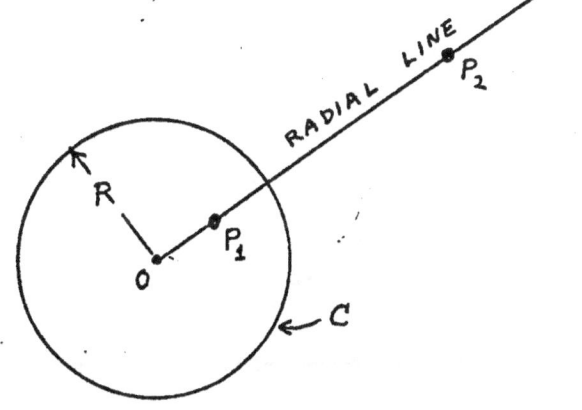

(i)   A radical line starting from the center   O   and extending to infinity

contains both P_1 and P_2.   The center   O   cannot be between P_1 and P_2.

(ii) $(\overline{O\,P_1}) \cdot (\overline{OP_2}) = R^2$, where R is the radius of the circle C.

(iii) If C is a straight line,

the points P_1 and P_2 are on

opposite sides of C. The line

segment $\overline{P_1\,P_2}$ is bisected by

C and is perpendicular to C.

Property III   Preservation of symmetric points

Suppose we have a circle C with two points P_1 and P_2 symmetric with

respect to it in the z-plane. The bilinear transformation (1) makes the

circle onto C′ and the points P_1 and P_2 onto $P_1′$ and $P_2′$ respectively in

the w-plane. The $P_1′$ and $P_2′$ are symmetric with respect to the circle C′.

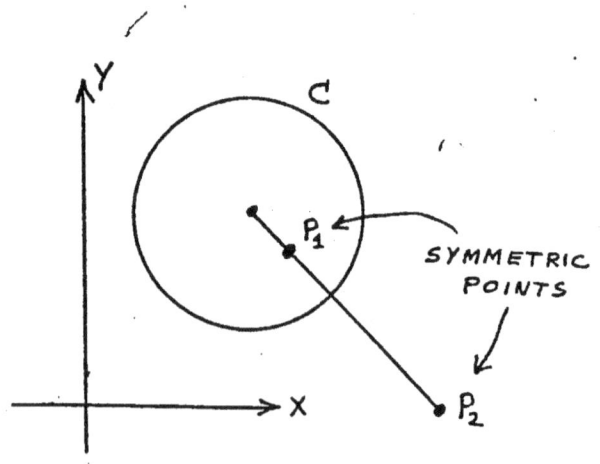
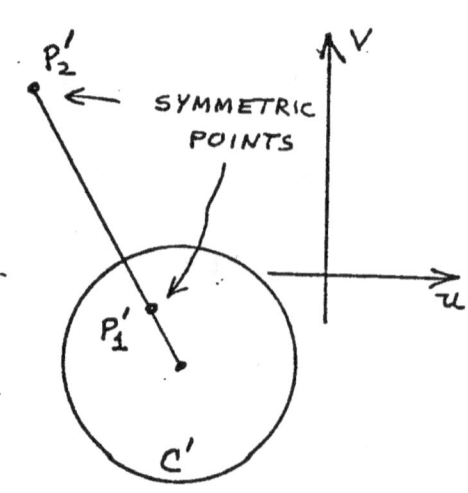

A method for proving Property III is outlined in supplementary problem

SP 9.10.8.

Example 3

Two points $1 + i$ and "q" are symmetric with respect to the circle $|z| = 2$.

Find "q".

Solution

Since $|1 + i| = \sqrt{2}$, then

$|q|$ must satisfy the relation

$\sqrt{2} \; |q| = 2^2$ which yields

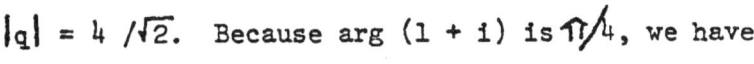

$|q| = 4 \, /\sqrt{2}$. Because arg $(1 + i)$ is $\pi/4$, we have

$$q = \frac{4}{\sqrt{2}} \; e^{i \; \pi/4} = 2(\sqrt{2} \; e^{i \; \pi/4})$$

$$q = 2(1 + i).$$

Thus the points $1 + i$ and $2 + 2i$ are symmetric with respect to the circle

$|z| = 2.$

Example 4

Let p and q be complex numbers, and suppose that the points in the

z-plane described by p and q are symmetric with respect to the circle

$|z - a| = R.$ Show that

(3) $q = a + \dfrac{R^2}{\overline{p} - \overline{a}}$.

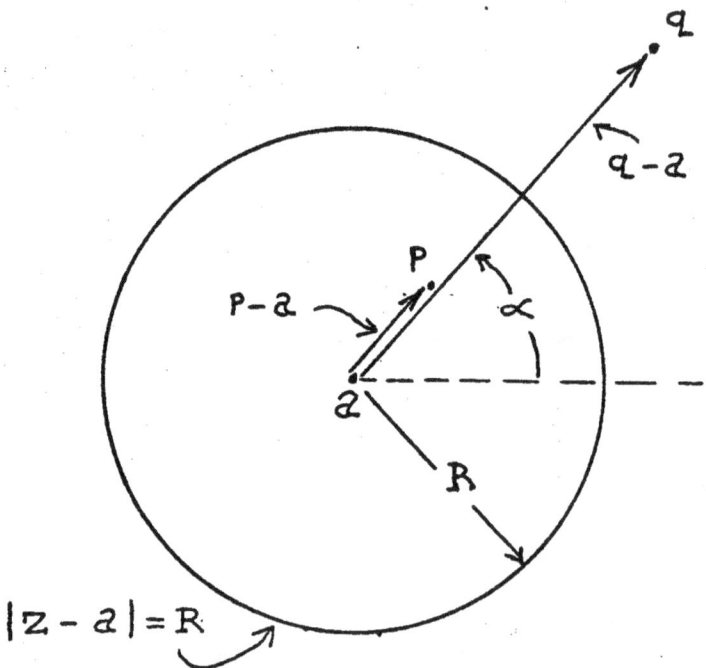

Solution

The firgure shows the

circle $|z - a| = R$ and

the two symmetric points

p and q. The condition

specifying the fact that p and q are symmetric is

$|p - a| \cdot |q - a| = R^2.$

Since

$$\overline{p - a} = |p - a| \, e^{-i\alpha} \qquad \text{and}$$

$$q - a = |q - a| \, e^{i\alpha}$$

we see that

$$R^2 = |p - a| \cdot |q - a| = (p - a)(q - a)$$

$$R^2 = (\overline{p} - \overline{a})(q - a)$$

and thus $\qquad q = a + \dfrac{R^2}{\overline{p} - \overline{a}}$.

Problems

● 10.2 Let $1 + \sqrt{3}\,i$ and q be points symmetric with respect to the circle

$|z| = 4$. Find q.

● 10.3 Let $7 + i$ and q be points symmetric with respect to the circle

$|z - 6| = 3$. Find q.

The fact that a bilinear transformation maps a circle C and two symmetric

points P_1 and P_2 into a circle C' with corresponding symmetric points

P_1' and P_2' can be used to find specific mapping functions as the following

example illustrates.

Example 4

Find a bilinear transformation mapping the upper half plane $\text{Re}\{z\} \geq 0$

onto the circle $|w| \leq 1$ and such that $z = 2i$ maps onto $w = 0$.

Solution

The points 2i and - 2i are

symmetric with respect to the

x-axis. These two points must

map into points symmetric with

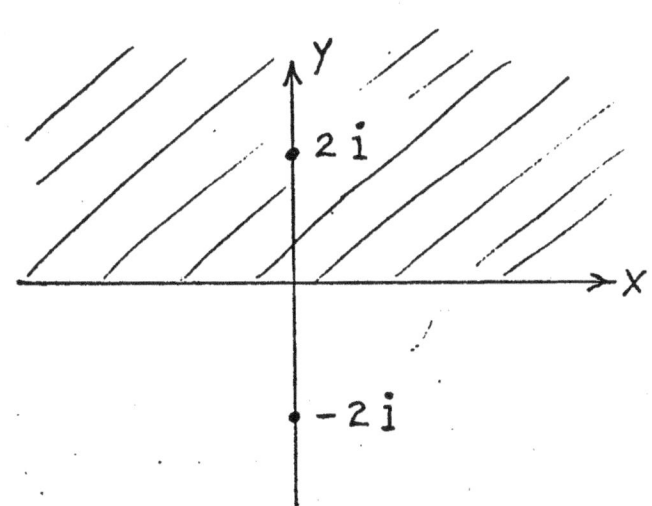

respect to the unit circle $|w| = 1$. The points 0 and ∞ are symmetric

with respect to $|w| = 1$, and since $z = 2i$ must map onto $w = 0$, then

$z = -2i$ must map onto $w = \infty$. The ratio

$$\omega = \frac{z - 2i}{z + 2i}$$

satisfies these requirements. Multiplication by a unit vector $e^{i\theta_0}$ will

not alter the size of the image circle and thus

$$w = e^{i\theta_0} \left(\frac{z - 2i}{z + 2i} \right)$$

maps as required.

Problems

10.4   Find a bilinear transformation mapping the half plane $Re(z) \geq 0$

onto the circle $|w| \leq 2$ and the point $z = 1$ onto $w = 0$.

10.5   Find a bilinear transformation mapping the half plane $y \geq x$ onto

the unit circle $|w| \leq 1$ and the point $z = i$ onto $w = 0$.

9.11 Poisson's integral formulas

Imagine the region shown represents

some uniform material and that the

temperature at each point Q on the

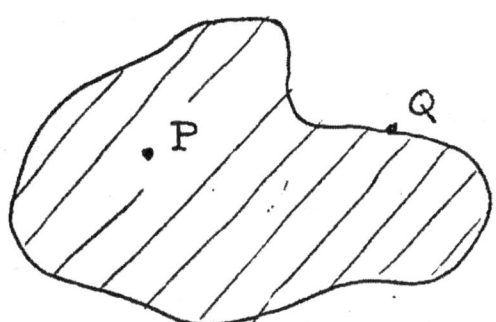

boundary is known. Suppose we wish to determine the steady state temperature at each point P inside the region. Special cases of this general problem were solved in previous sections. This problem is called the Dirichlet problem.

The Dirichlet Problem:

Let the values of a harmonic function be given on the boundary of some simply connected region. Determine the values of the function inside the region.

It can be shown that a unique solution to the Dirichlet problem exists under very loose restrictions on the shape of the region and the boundary values.

There are two cases in which a simple formula can be written for the solution of the Dirichlet problem. The first is the case where the region is the unit circle $|z| \leq 1$, and the second is the case where the region is the upper half plane $\text{Im}(z) \geq 0$.

In the first case we have

$$(1) \quad u(r, \theta) = \frac{1}{2\pi} \int_0^{2\pi} \frac{(1 - r^2)\, u\,(1,\, \phi)\, d\phi}{1 - 2r \cos(\theta - \phi) + r^2}$$

which is called Poisson's formula for a circle. Here (r, θ) are the polar

coordinates of a point in the unit circle $|z| \leq 1$ and $(1, \emptyset)$ is a general

point on the boundary.

The function $u(1, \emptyset)$ is

assumed known, and the

right hand side of (1)

allows us to compute the

harmonic function $u(r, \theta)$

at the point (r, θ) inside

the circle by simply knowing

its values $u(1, \emptyset)$ on the

boundary.

Our second formula is

(2) $\quad u(x, y) = \dfrac{1}{\pi} \displaystyle\int_{-\infty}^{\infty} \dfrac{u(\gamma, 0)\, y\, d\gamma}{(\gamma - x)^2 + y^2}$

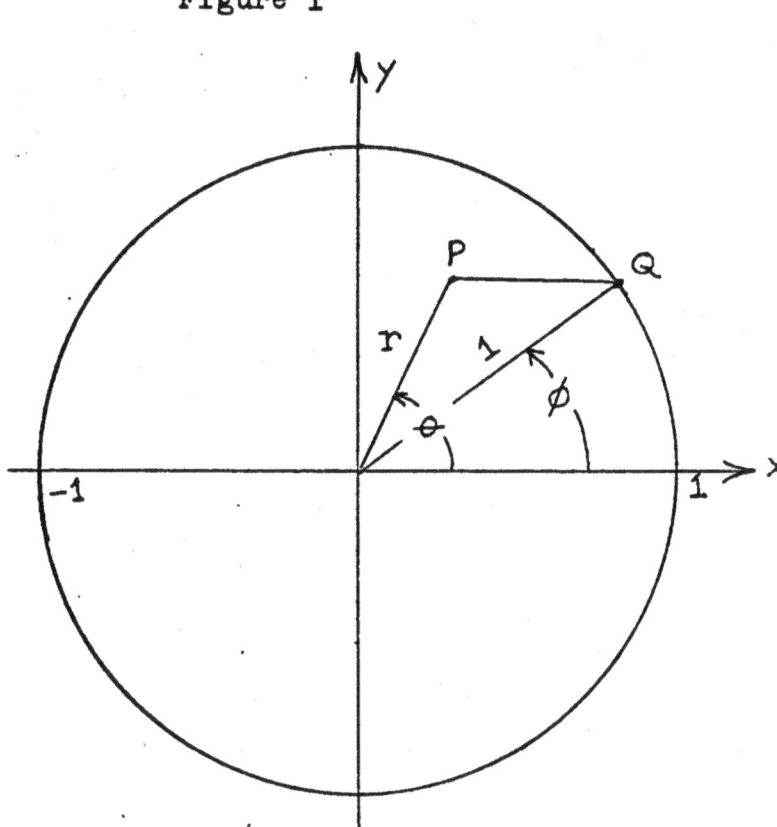

Figure 1

which is called Poisson's formula for a half plane. Here the values of

the harmonic function u are assumed known on the real axis $(u(\gamma, 0))$,

and the right hand side of (2) allows us to compute $u(x, y)$ at any point

(x, y) in the upper half plane $y > 0$ by simply knowing the values of u

on the x-axis.

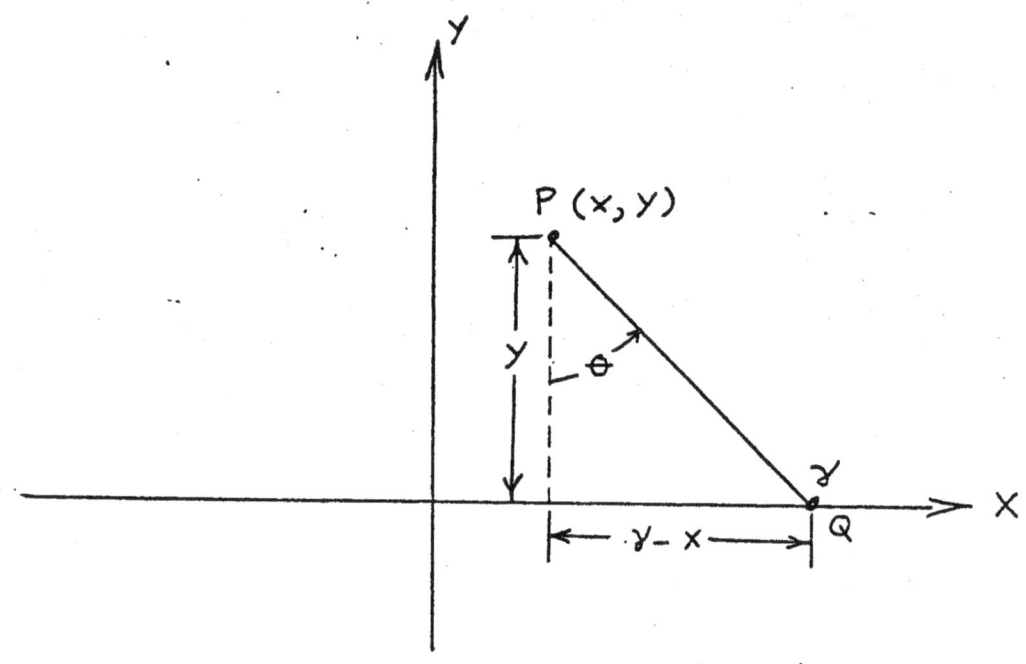

Figure 2

It is useful to determine the geometric significance of the terms that appear in the integrands of (1) and (2). Looking at Figure 1 we see that

$$|\overline{PQ}|^2 = 1 + r^2 - 2r \cos (\theta - \emptyset)$$

from the law of cosines. Thus (1) can be written as

$$(3) \quad u(r, \theta) = \frac{1 - r^2}{2\pi} \int_0^{2\pi} \frac{u(1, \emptyset) \, d\emptyset}{|\overline{PQ}|^2}$$

In (3), the point P at (r, θ) is fixed, while the point Q at $(1, \emptyset)$ moves over the boundary of the circle to perform the integration.

To see the geometric meaning in (2), we examine Figure 2. Here the point

P at (x, y) is fixed, while the point Q moves from left to right over the x-axis to perform the integration. Thus x and y are thought of as constants while γ is variable. Manipulating the integrand we have

$$\frac{y \, d\gamma}{(\gamma - x)^2 + y^2} = \frac{\dfrac{d\gamma}{y}}{1 + \left(\dfrac{\gamma - x}{y}\right)^2}$$

$$= \frac{d\left(\dfrac{\gamma - x}{y}\right)}{1 + \left(\dfrac{\gamma - x}{y}\right)^2} \, .$$

From Figure 2 we see that

$$\theta = \tan^{-1}\left(\frac{\gamma - x}{y}\right) .$$

Recalling from the elementary calculus that

$$d \tan^{-1} u = \frac{du}{1 + u^2}$$

we see that $\left(u = \dfrac{\gamma - x}{y}\right)$

$$\frac{y \, d\gamma}{(\gamma - x)^2 + y^2} = d \tan^{-1}\left(\frac{\gamma - x}{y}\right)$$

$$= d\theta .$$

Thus we can write (2) as

$$(4) \quad u(x, y) = \frac{1}{\pi} \int_{\theta = -\frac{\pi}{2}}^{\theta = \frac{\pi}{2}} u(\gamma, 0) \, d\theta$$

Before trying to derive the Poisson integrals, it is instructive to use them to actually calculate the numerical values of a harmonic function.

Example 1

Suppose the temperature is measured at points on the boundary of the

unit circle as shown in the table. Estimate the temperature at $z = \frac{1}{2}$

using Poisson's integral (1).

angle ϕ in degrees	temperature $u(1, \phi)$ on the boundary
0°	0
30°	10
60°	20
90°	30
120°	20
150°	10
180°	0
·	·
·	·
·	·
330°	0

Solution

To estimate the value of the

temperature u at the point ($\frac{1}{2}$, 0), we

will approximate the integral (3) by a

Riemann sum. Since there are twelve

points on the unit circle where temperatures

are measured, it is convenient to use twelve

terms for the Riemann sum. In figure 3,

we see the boundary of the unit circle

subdivided into twelve equal parts.

We substitute for $d\phi$ 30° or $\pi/6$ radians. We can measure the distances

$\overline{PQ}$ from Figure 3.

We have then

$$u(r, \theta) = \frac{1 - r^2}{2\pi} \int_0^{2\pi} \frac{u(1, \phi)\, d\phi}{|PQ|^2}$$

$$u(\tfrac{1}{2}, 0) = \frac{.75}{2\pi} \int_0^{2\pi} \frac{u(1, \phi)\, d\phi}{|PQ|^2}$$

$$u(\tfrac{1}{2}, 0) \approx \frac{.75}{2\pi} \left[\frac{u(1, 0)}{|PQ|^2}\Big|_{\phi=0^\circ} \cdot \left(\frac{\pi}{6}\right) + \frac{u(1, 30^\circ)}{|PQ|^2}\Big|_{\phi = 30^\circ} \cdot \left(\frac{\pi}{6}\right) \right.$$

$$\left. + \frac{u(1, 60^\circ)}{|PQ|^2}\Big|_{\phi=60^\circ} \cdot \left(\frac{\pi}{6}\right) + \cdots + \frac{u(1, 330^\circ)}{|PQ|^2}\Big|_{\phi = 330^\circ} \cdot \left(\frac{\pi}{6}\right) \right]$$

$$u(\tfrac{1}{2}, 0) \approx \frac{.75\pi}{2\pi 6} \left[\frac{0}{(.5)^2} + \frac{10}{(.62)^2} + \frac{20}{(.87)^2} + \right.$$

$$\frac{30}{(1.12)^2} + \frac{20}{(1.34)^2} + \frac{10}{(1.47)^2} + \frac{0}{(1.5)^2} + 0 + 0 \cdots + 0 \left. \right]$$

$$u(\tfrac{1}{2}, 0) \approx .0625 \left[26.01 + 26.42 + 23.92 + 11.14 + 4.63 \right]$$

$$u(\tfrac{1}{2}, 0) \approx 5.76.$$

Of course, in computing this approximation of the temperature at $(\tfrac{1}{2}, 0)$

we have assumed that the temperature distribution is not wildly varying

between the points $\phi = 0^\circ$, 30°, 60°, $\cdots$. For example, we might assume

that $u(1, \phi)$ varies according to the graph shown.

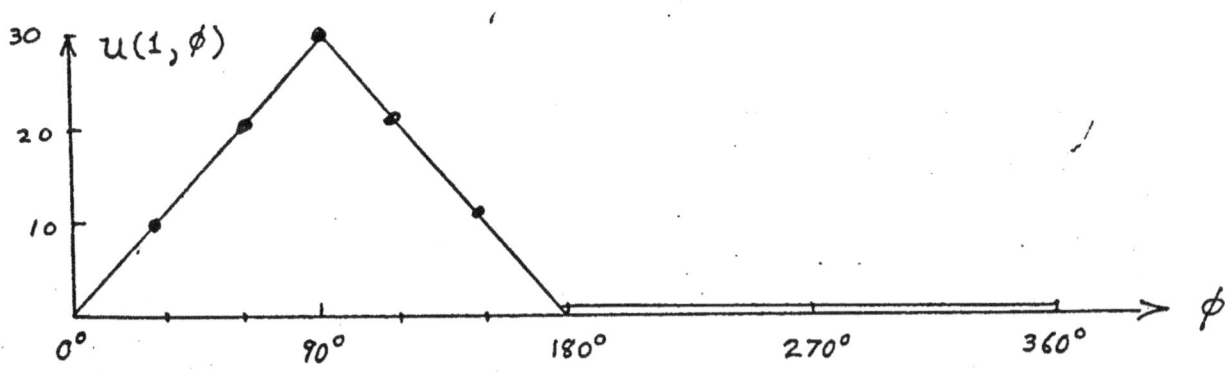

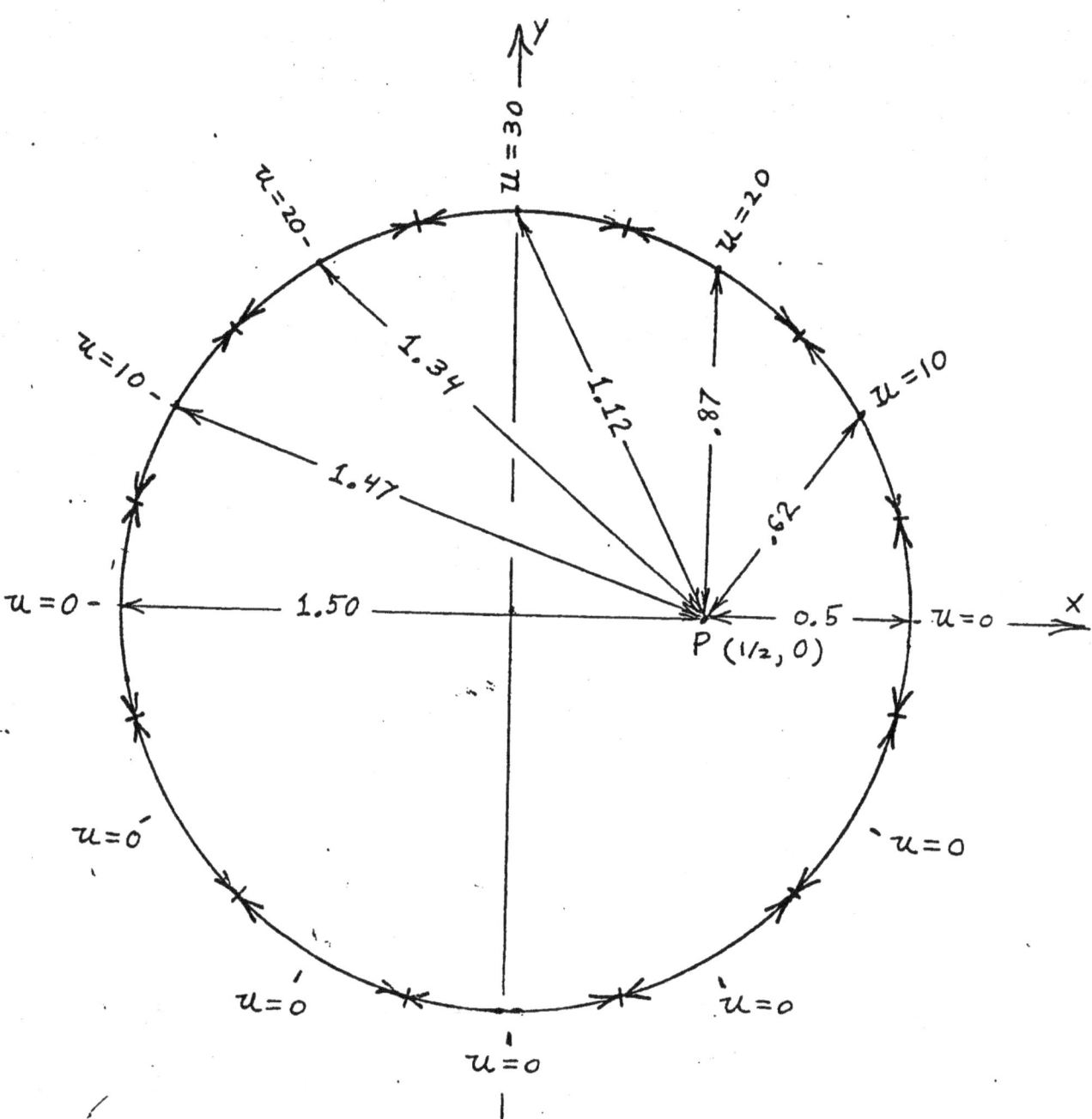

Figure 3

Problems

● **11.1** The values of the temperature on the boundary of the unit circle are measured at $\phi = 5^\circ, 15^\circ, 25^\circ, 35^\circ, \cdots, 175^\circ$ and found in each case to be 20° centigrade. However, for $\phi = -5^\circ, -15^\circ, -25^\circ, \cdots, -175^\circ$ the temperature is found to be zero degrees centigrade in each case. Estimate the temperature at the point $x = 0, y = \frac{1}{2}$. Compare your answer with the solution to Example 4 of section 9.9.

● **11.2** Use the Poisson integral (3) to estimate the temperature at $(0, 0)$ required in problem 46 of this chapter. Measure the temperature on the boundary $u(1, \phi)$ at 360 evenly spaced points $\phi = 0^\circ, 1^\circ, 2^\circ, \cdots, 359^\circ$. Compare your estimate with the exact solution obtained previously.

● **11.3** Prove that the value of a harmonic function at the point $z = 0$ equals the average of its values on the circular boundary $|z| = 1$.

Next we turn our attention to Poisson's formula for the half plane (2) and its equivalent (4).

Example 2

Estimate the temperature at the point $z = i$, when the temperatures on the x-axis are found to be

$$u(x, 0)= \begin{cases} 0 \text{ for } x < 1 \\ 10x \text{ for } 1 \leq x \leq 3 \\ 0 \text{ for } 3 < x. \end{cases}$$

Solution

We will use Poisson's integral (4).

$$u(x, y) = \frac{1}{\pi} \int_{\theta=-\pi/2}^{\pi/2} u(\gamma, 0) \, d\theta$$

$$u(0, 1) = \frac{1}{\pi} \int_{\theta=\pi/4}^{\theta=tan^{-1}3} 10\gamma \, d\theta . = \frac{1}{\pi} \int_{\theta=\pi/4}^{\theta=tan^{-1}3} 10\gamma \, d(tan^{-1}\gamma)$$

This last integral is true because u(γ , 0) = 0 for values of θ outside

the range $45° \leq 0 \leq tan^{-1} 3 = 71.6°$.

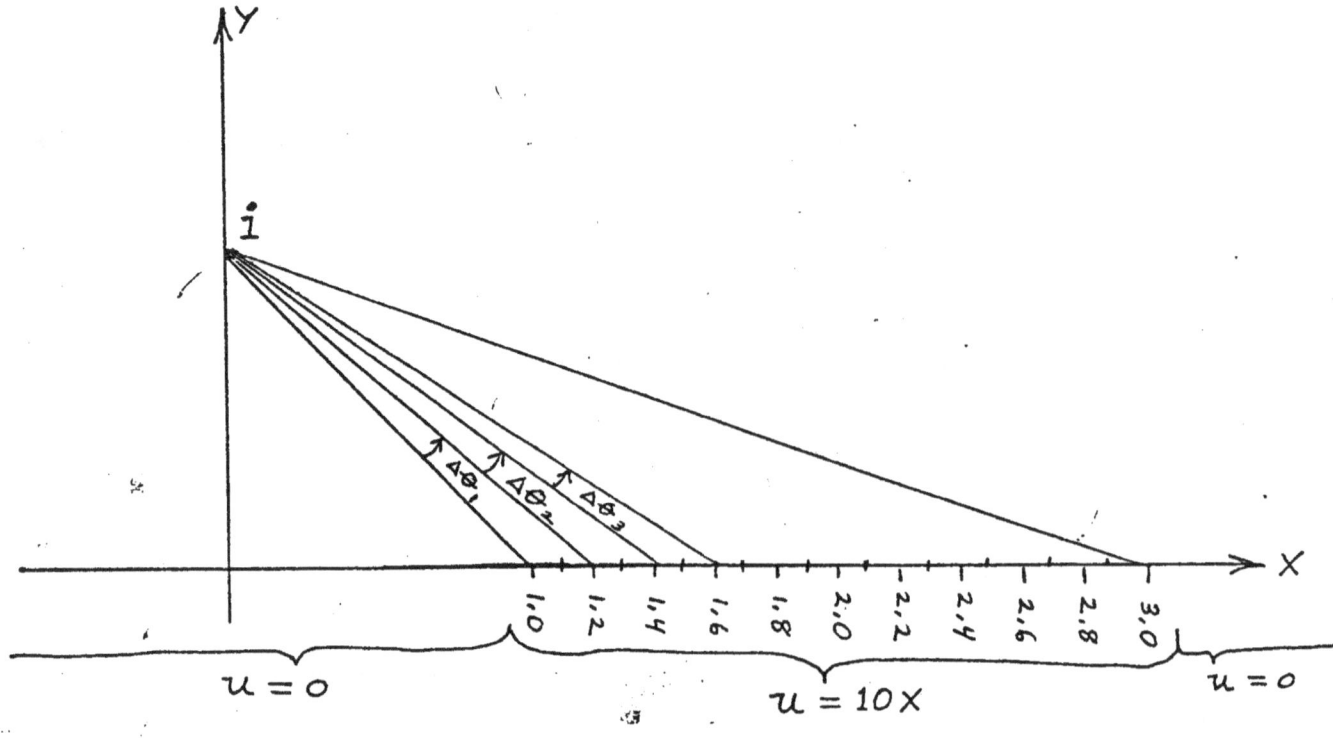

For convenience, we subdivide the interval $1 \le x \le 3$ into 10 equal sub-intervals of length 0.2. With this subdivision we generate an approximation by means of a Riemann sum for the last integral. We see that the end points of the intervals are 1, 1.2, 1.4, $\cdots$, 2.8, 3.0. We select the value of u at the midpoint of each interval. Thus we will select u at x = 1.1, 1.3, 1.5, $\cdots$, 2.9, for the values of the temperature on the boundary.

We get $u(0, 1) \approx \sum_{k=1}^{10} 10x_k \dfrac{\Delta\theta_k}{\pi}$ (angle in radians)

where $x_1 = 1.1$, $x_2 = 1.3$, $\cdots$, $x_{10} = 2.9$,

and where

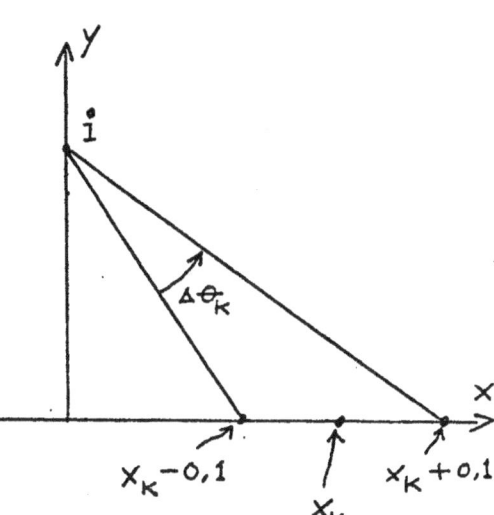

$$\dfrac{\Delta\theta_k \text{ (radians)}}{\pi} = \dfrac{\Delta\theta_k \text{ (degrees)}}{180}$$

$$= \dfrac{\tan^{-1}(x_k + 0.1) - \tan^{-1}(x_k - 0.1)}{180} \quad \text{(degrees)}$$

We summarize our calculations in tabular form:

k	x_k	$u(x_k, 0) =$ $10x_k$	$\Delta\theta_k$ (degrees)	$10x_k \; \Delta\theta_k$ (degrees)
1	1.1	11	5.2°	57.2
2	1.3	13	4.3°	55.9
3	1.5	15	3.5°	52.5
4	1.7	17	3.0°	51
5	1.9	19	2.5°	47.5
6	2.1	21	2.1°	44.1
7	2.3	23	1.8°	41.4
8	2.5	25	1.6°	40
9	2.7	27	1.4°	37.8
10	2.9	29	1.2°	34.8
				462.2

Thus we have

$$u(0, 1) \approx \sum_{k=1}^{10} 10x_k \; \frac{\Delta\theta_k \; (\text{degrees})}{180}$$

$$= \frac{462.2}{180} = 2.57$$

Example 3

Find an approximate value for the temperature considered in the previous

example by approximating Poisson's integral with equal segments of angle

$d\theta = 10^\circ$.

Solution

In the figure we see radial lines emerging from $z = i$ making angles $\theta = 0^\circ, \pm 10^\circ, \pm 20^\circ, \cdots, \pm 90^\circ$ with the negative y-axis. We see that only the rays making angles $\theta = 50^\circ$, 60° and 70° meet the x-axis in the interval $1 \leq x \leq 3$ on which the temperature u is not zero.

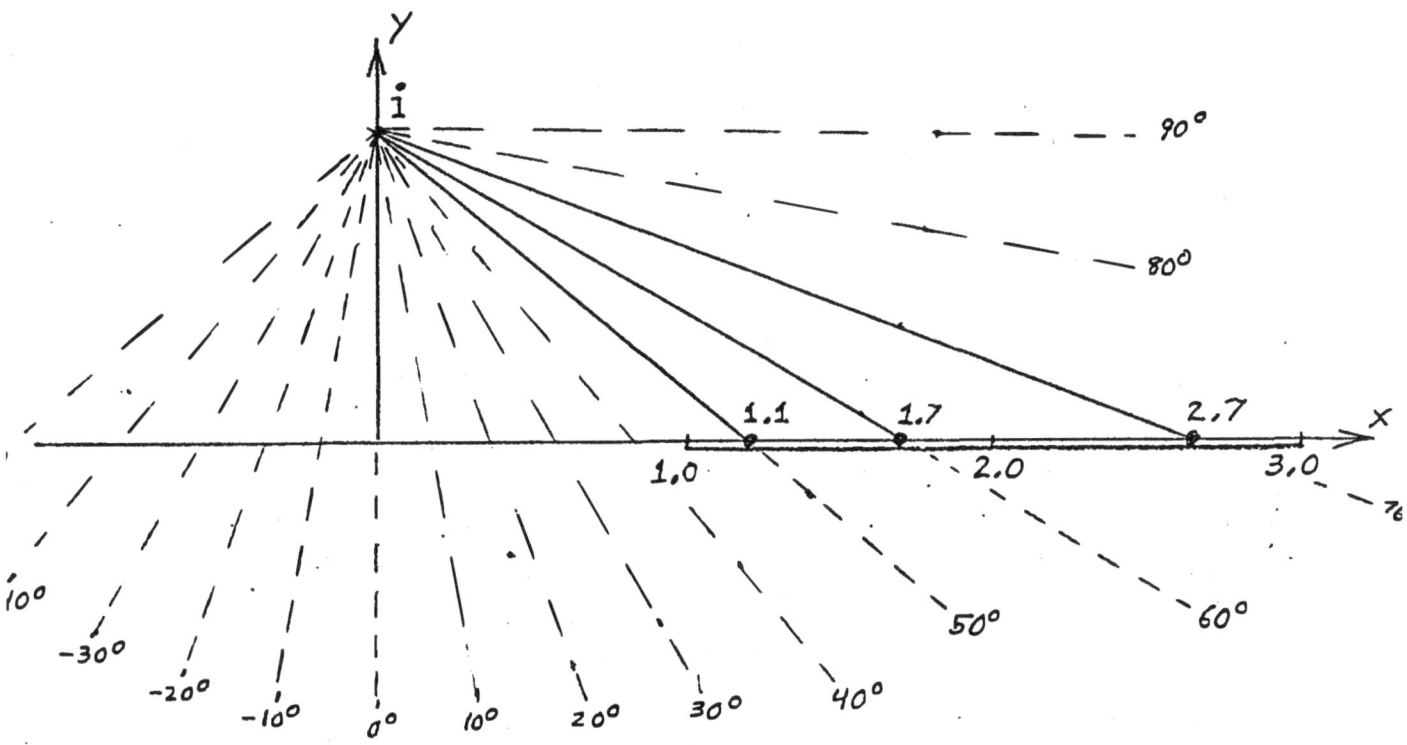

These three rays meet the x-axis at the points where x is 1.1, 1.7 and 2.7 and thus the temperatures (10x) there are 11, 17 and 27 respectively. Since $d\theta$ is 10° we have $\dfrac{d\theta \text{ (radians)}}{\pi} = \dfrac{10^\circ}{180^\circ} = \dfrac{1}{18}$

Thus our approximation is

$$u(0, 1) = \int_{0 = -\frac{\pi}{2}}^{\frac{\pi}{2}} u(\gamma, 0) \ \frac{d\theta}{\pi}$$

$$\approx \sum_{k} u(\gamma_k, 0) \ \frac{d\theta_k}{\pi}$$

$$\approx 11(1/18) + 17(1/18) + 27(1/18)$$

$$\approx 3.06.$$

This answer is 20% larger than the approximation found in Example 2. Since

the previous approximation involved 10 selected values of the temperature

and the present only three, the previous answer, 2.57, is probably more

accurate.

Example 4

Use Poisson's integral (4) to find the exact bounded temperature at any

point (x, y) in the upper half plane when the values on the x-axis are given

by

$$u(x, 0) = \begin{cases} 3 & \text{for } 0 < x \\ -2 & \text{for } x < 0 . \end{cases}$$

<u>Solution:</u>

.We have

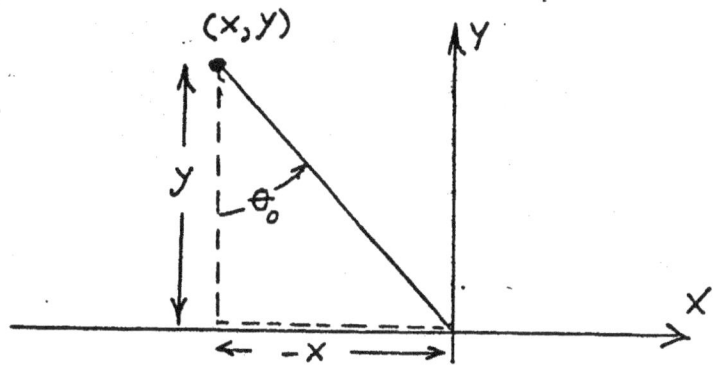

$$u(x, y) = \frac{1}{\pi} \int_{\theta = -\pi/2}^{\pi/2} u(\gamma, 0) \, d\theta$$

$$u(x, y) = \frac{-2}{\pi} \int_{-\pi/2}^{\theta_0} d\theta + \frac{3}{\pi} \int_{\theta_0}^{\pi/2} d\theta$$

where $\theta_0 = \tan^{-1} \frac{-x}{y} = -\tan^{-1} \frac{x}{y}$.

Thus we have

$$u(x, y) = \frac{-2}{\pi} (\theta_0 + \pi/2) + (3/\pi)(\pi/2 - \theta_0)$$

$$u(x, y) = \frac{1}{2} - (5/\pi) \theta_0$$

$$u(x, y) = \frac{1}{2} + (5/\pi) \tan^{-1} x/y .$$

Since $\tan^{-1} \frac{x}{y} = \frac{\pi}{2} - \tan^{-1} \frac{y}{x}$.

$$u(x, y) = 3 - (5/2) \tan^{-1} y/x .$$

This same problem was solved by another method in Example 1 of section 9.7.

Problems

11.4 Consider the temperature distribution given in Example 2. Use the

method of that example to approximate the temperature at the point $z = 1 + i$.

Use the same values of x_k.

11.5 Get a second approximation to the temperature at $z = 1 + i$ found in the

previous problem. Use the method of Example 3 with $d\theta = 10^{\circ}$.

In problems 61 and 62, use Poisson's integral formula to find the values

of a bounded harmonic function $u(x, y)$ in the upper half of the z-plane

when the values $u(x, 0)$ are given on the x-axis.

11.6
$$u(x, 0) = \begin{cases} 5 & \text{for } 1 < x \\ 7 & \text{for } -1 < x < 1 \\ -10 & \text{for } x < -1 . \end{cases}$$

Compare your answer with Example 2 of section 9.7.

11.7
$$u(x, 0) \begin{cases} 2 & \text{for } 5 < x \\ 0 & \text{for } 0 < x < 5 \\ 2 & \text{for } x < 0 \end{cases}$$

(See problem 20(c) of this chapter.)

We will now derive Poisson's half plane formula (2). Consider the contour

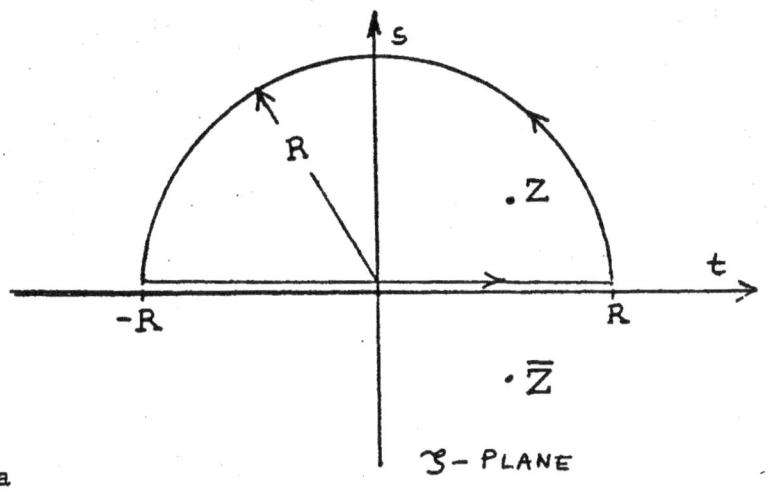

$\mathcal{J}$ - PLANE

C_r shown. Let $f(z)$ be a function analytic for z in the upper half plane. We have from Cauchy's integral formula

$$f(z) = \frac{1}{2\pi i} \oint \frac{f(\mathcal{J})\, d\mathcal{J}}{\mathcal{J} - z}$$

for z inside C_r. Since $\bar{z}$ is outside this contour we have

$$0 = \frac{-1}{2\pi i} \oint \frac{f(\mathcal{J})\, d\mathcal{J}}{\mathcal{J} - \bar{z}}$$

Adding these last two expressions we get

$$f(z) - 0 = \frac{1}{2\pi i} \oint \frac{f(\mathcal{J})\, d\mathcal{J}}{\mathcal{J} - z} - \frac{1}{2\pi i} \oint \frac{f(\mathcal{J})\, d\mathcal{J}}{\mathcal{J} - \bar{z}}$$

$$f(z) = \frac{1}{2\pi i} \oint f(\mathcal{J}) \left[\frac{1}{\mathcal{J} - z} - \frac{1}{\mathcal{J} - \bar{z}} \right] d\mathcal{J}$$

$$f(z) = \frac{1}{2\pi i} \oint \frac{f(\mathcal{J})\, (z - \bar{z})\, d\mathcal{J}}{(\mathcal{J} - z)(\mathcal{J} - \bar{z})}.$$

Now write

$$f(z) = \frac{1}{2\pi i} \int_{\longrightarrow} + \frac{1}{2\pi i} \int_{\frown}$$

and let $R \to \infty$. We assume that $f(z)$ is such that $\displaystyle\lim_{R \to \infty} \int_{\frown} = 0$. If $|f(z)|$ is bounded for $\text{Im}(z) \geq 0$ this limit will vanish as required. (See

Theorem 1 of section 6.7). On the integral $\int$ the variable $\mathcal{Y}$ of

integration is on the real axis and thus $\mathcal{Y} = t$. We have

$$f(z) = \frac{1}{2\pi i} \int_{-\infty}^{\infty} \frac{f(t)\ 2y\ i\ dt}{(t - x - iy)(t - x + i y)}$$

$$f(z) = \frac{1}{\pi} \int_{-\infty}^{\infty} \frac{f(t)\ y\ dt}{(t - x)^2 + y^2} \ .$$

Next write $f(z) = u(x, y) + i v(x, y)$ and get

$$u(x, y) + i v (x, y) = \frac{1}{\pi} \int_{-\infty}^{\infty} \frac{u(t, 0)\ y\ dt}{(t - x)^2 + y^2} + \frac{i}{\pi} \int_{-\infty}^{\infty} \frac{v(t,0)\ y\,dt}{(t-x)^2 + y^2}$$

Taking the real part of this last expression we get the required Poisson

integral

$$u(x, y) = \frac{1}{\pi} \int_{-\infty}^{\infty} \frac{u(t, 0)\ y\ dt}{(t - x)^2 + y^2} \ .$$

A method for deriving Poisson's formula (1) for the circle is outlined

in supplementary problem 9.11.7.

The Poisson formulas solve the Dirichlet problem when the region is a

circle or a half plane. How can we solve the Dirichlet problem for other

regions? If we can find an analytic function $f(z)$ which maps the region

of interest R onto the unit circle $|w| \le 1$, then we can map the boundary

values of a harmonic function on R onto the boundary $|w| = 1$ also. Solve

the resulting Dirichlet problem on the circle with Poisson's integral and then

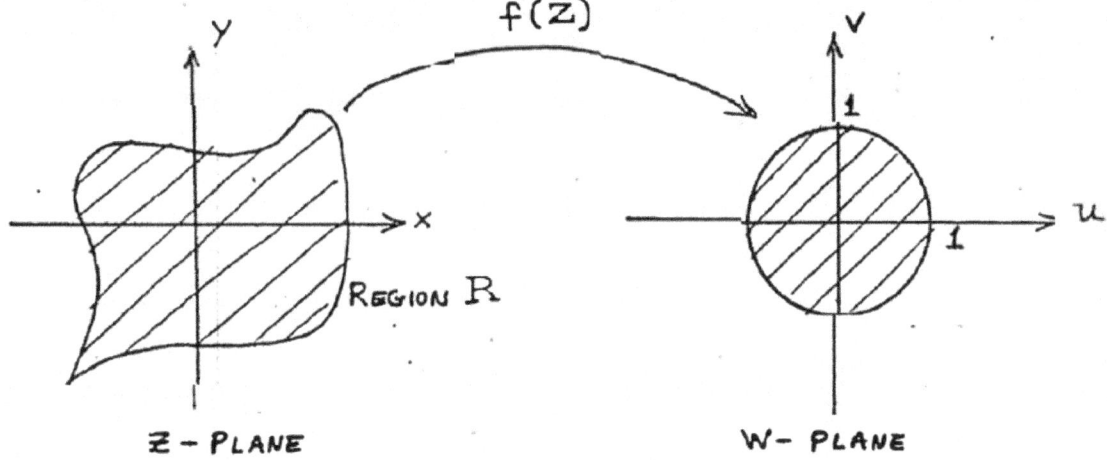

map the solution back to R in the z-plane with the function $z = f^{-1}(w)$.

Thus we see that the Dirichlet problem can be solved for the region R if

we can find a conformal mapping $w = f(z)$ of R onto $|w| \leq 1$. Can the

function $f(z)$ always be found? Unfortunately there is no general method

for constructing $f(z)$ given R. However, there is a result known as the

Riemann Mapping Theorem which states that if R is simply connected, then

$f(z)$ exists. While we might be frustrated in our attempts to find or

construct a specific $f(z)$, we can take some comfort, at least, in knowing

that it does exist and that effort might result in finding it.

<u>9.12 The Schwarz – Christoffel transformation</u>

In this section we examine a formula which enables us to map the upper half of the z-plane onto the interior of a general polygon in the w-plane.

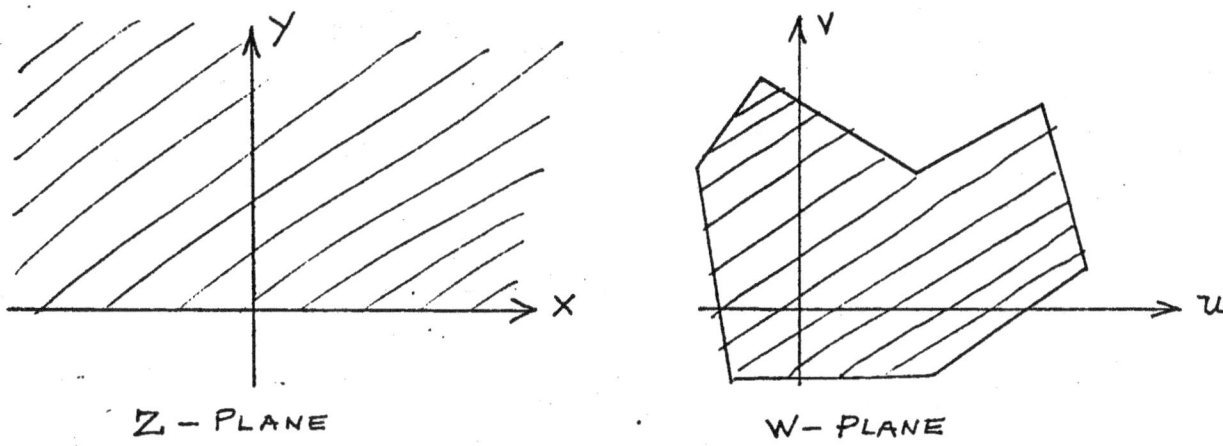

Z - PLANE   W - PLANE

Of course, the boundary of the polygon will map onto the x-axis. Before

describing the formula which describes this mapping we must specify con-

ventions regarding the vertices, angles, and orientation of the polygon.

Notations and Conventions Regarding Polygons

(i) Select a point w_∞ on the

boundary of the polygon which

will map onto the point $z = \infty$.

This point may, or may not be

a vertex of the polygon.

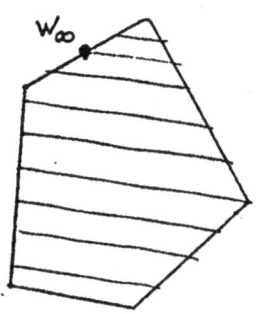

(ii) The positive orientation

of the boundary of the polygon

is the same as the positive

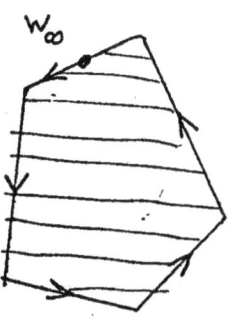

orientation for contours of

integration. In the case at

hand it is clockwise. In

more complicated cases we

imagine walking on the

boundary so that the interior

(shaded) is always on our left.

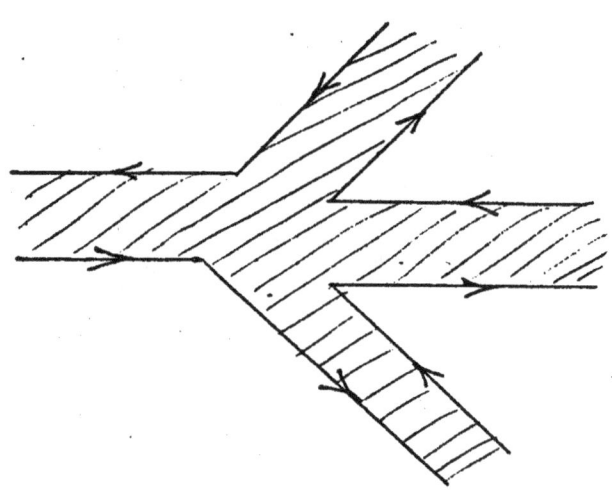

(iii) Starting from w_∞ , traverse the boundary of the polygon in the

positive sense naming the vertices in succession w_1, w_2, $\cdots$, w_n. These

will map onto points x_1, x_2, $\cdots$, x_n, respectively, of the x-axis in the

z-plane which are in increasing order $x_1 < x_2 < \cdots < x_n$.

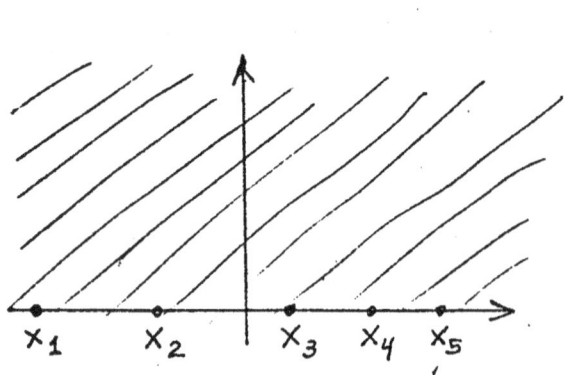

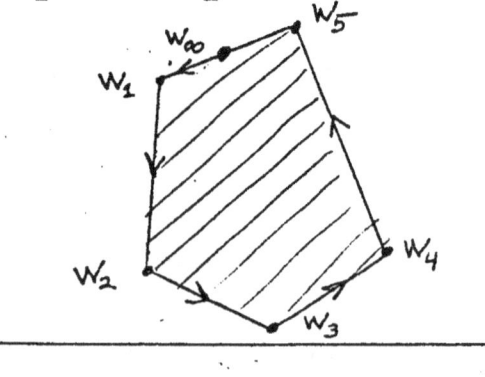

In this case, if $w_\infty = w_5$, then only the four points x_1, x_2, x_3, x_4 would

be shown since x_5 is now at infinity.

(iv) In this example, as we

move from w_∞ through w_1

to w_2 we make an abrupt

change in direction at w_1.

This change in direction,

called the "exterior angle

at w_1" is denoted by "$\alpha_1 \pi$".

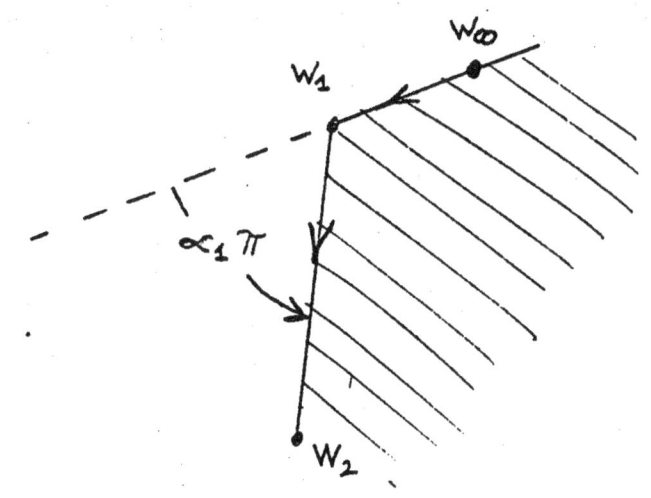

It is measured, as usual, in the positive or counterclockwise sense.

The exterior angles at

each vertex are

measured and denoted

by $\alpha_1 \pi$, $\alpha_2 \pi$, $\cdots$,

$\alpha_n \pi$.

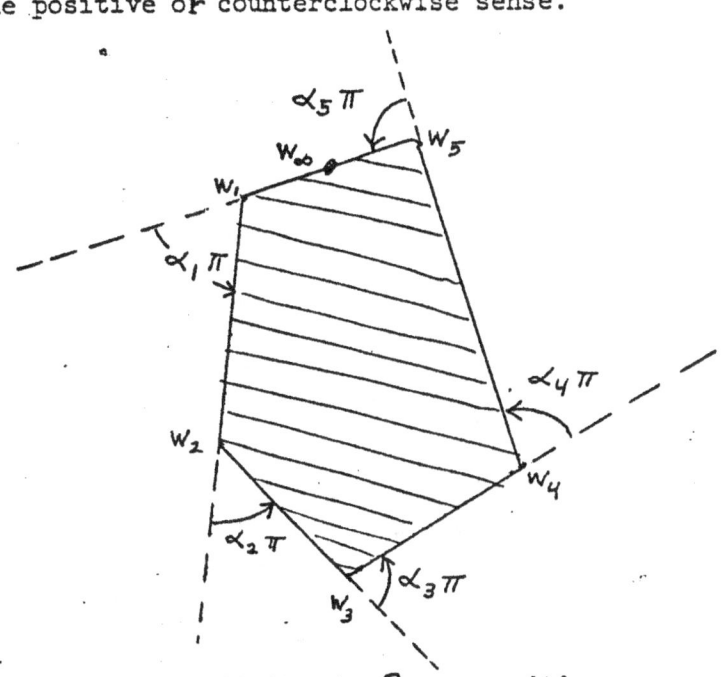

Remarks:

1. In the case of the polygon illustrated, all the $\alpha_i \pi$ are positive.

However, negative exterior angles

can occur as at the corner illustrated

where we must measure in the clockwise

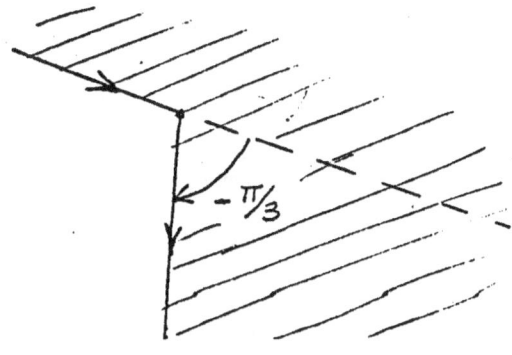

sense.

2. The sum of the exterior angles of a closed polygon must equal 2π.

$$\alpha_1\pi + \alpha_2\pi + \cdots + \alpha_n\pi = 2\pi.$$

Thus we have

(1)
$$\boxed{\alpha_1 + \alpha_2 + \cdots + \alpha_n = 2.}$$

We can now write the formula which maps the upper half of the z-plane

onto the polygon in the w-plane known as the Schwarz - Christoffel transformation:

Schwarz-Christoffel Transformation

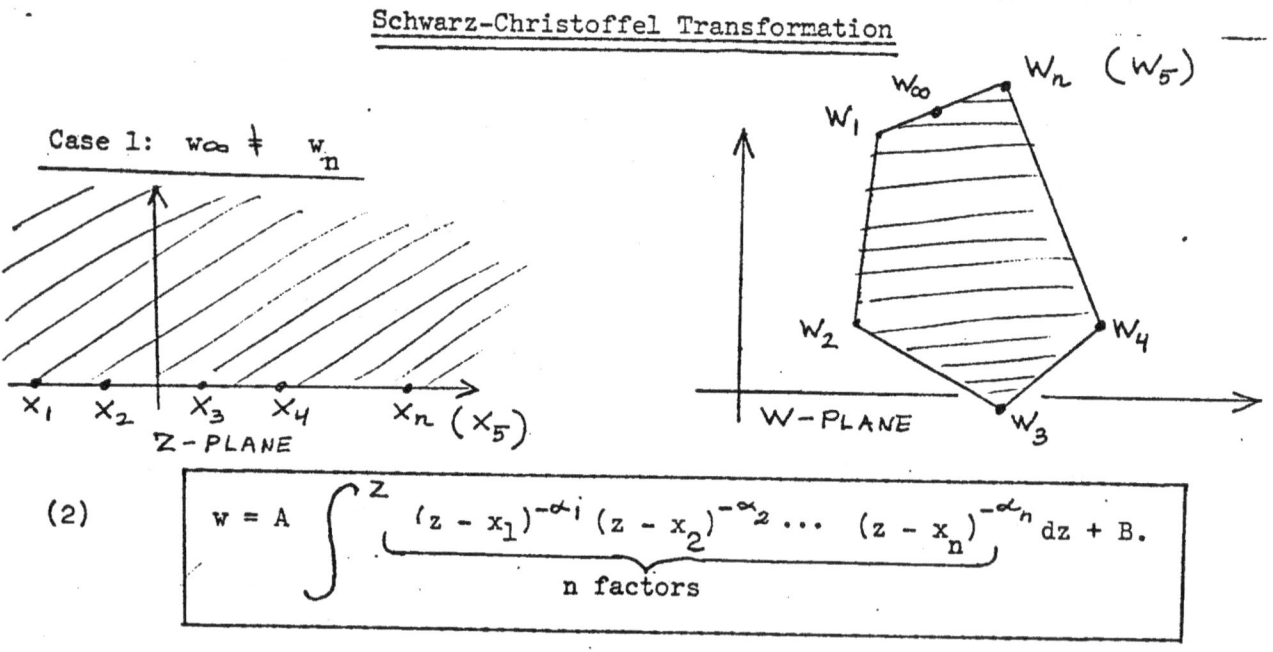

Case 1: $w_\infty \neq w_n$

Z-PLANE

W-PLANE

(2)
$$w = A \int^z \underbrace{(z - x_1)^{-\alpha_1}(z - x_2)^{-\alpha_2} \cdots (z - x_n)^{-\alpha_n}}_{n\ \text{factors}} dz + B.$$

Case 2: $w_\infty = w_n$

Now $x_n = \infty$. In this case we simply drop the factor $(z - x_n)^{-\alpha_n}$ from

the above formula and get only $n - 1$ factors in the integrand:

(3)
$$w = A \int^z (z - x_1)^{-\alpha_1} (z - x_2)^{-\alpha_2} \cdots (z - x_{n-1})^{-\alpha_{n-1}} \, dz \; + B$$

Remarks (continued)

3. We can picture the mapping

effected by (2) or (3) by

imagining the polygon in

the w-plane as a sheet of

rubber.

We break the rubber at w

and begin to stretch the

polygon open.

Continue opening the polygon

until   the boundary is a

straight line.

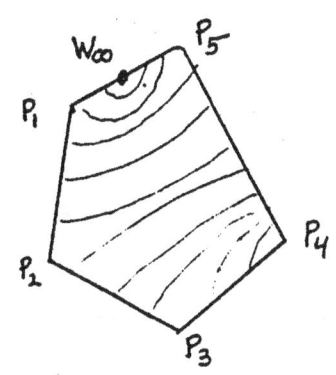

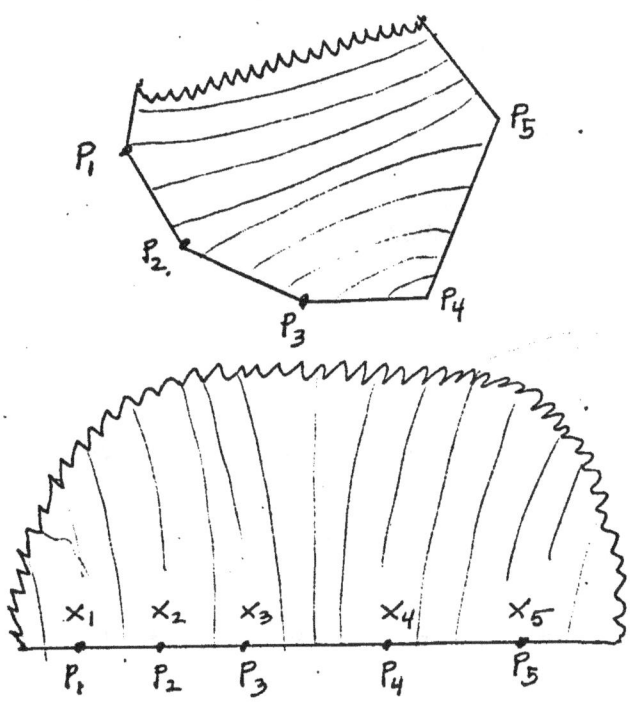

The points P_1, P_2, $\cdots$, P_5, must finally correspond to x_1, x_2, $\cdots$, x_5

on the x-axis, while the ragged edge resulting from the initial break at

w_∞ is stretched to infinity. The rubber sheet is now the upper half of the z-plane.

4. The constants A and B are in general complex numbers. By appropriately selecting these numbers we can effect a translation (B), magnification (|A|) and rotation (arg A) of the polygon.

5. When the vertices w_1, w_2, $\cdots$, w_n are given points in the w-plane, their images x_1, x_2, $\cdots$, x_n on the x-axis cannot be selected in an arbitrary manner. We can <u>arbitrarily select three</u> of these x_i's, but the remainder must then be calculated. In general, this calculation can be very difficult.

A method for deriving (2) is outlined in supplementary problem 9.12.21.

<u>Example 1</u>

Without specifying the numbers A, B or the x_i's, write both forms of the Schwarz-Christoffel transformation, (2) and (3), for the mapping of the upper half of the z-plane onto a rectangle in the w-plane.

<u>Solution</u>

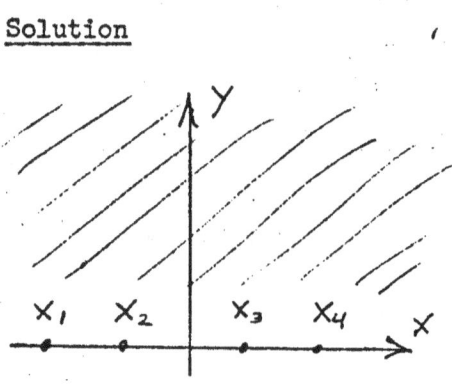

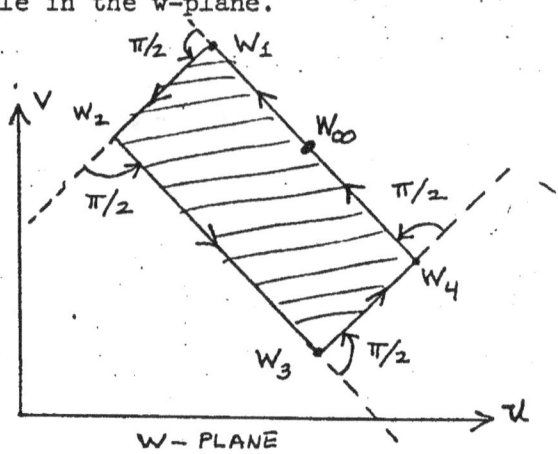

Notice that all the exterior angles are $\pi/2$, and thus $\alpha_i = \frac{1}{2}$ for i =

1, 2, 3, 4. If $w_\infty \neq w_4$ we have from (2)

$$w = A \int^z \frac{dz}{\sqrt{(z - x_1)(z - x_2)(z - x_3)(z - x_4)}} + B$$

If $w_\infty = w_4$, then the point x_4 is at infinity and we use (3) to get

$$w = A \int^z \frac{dz}{\sqrt{(z - x_1)(z - x_2)(z - x_3)}} + B .$$

In many problems of interest, one or more vertices of the polygon are

at infinity as the next example illustrates.

Example 2

Without specifying A, B,

or the x_i's, write both forms

of the Schwarz-Christoffel

transformation mapping the

upper half of the z-plane

onto the polygon shown.

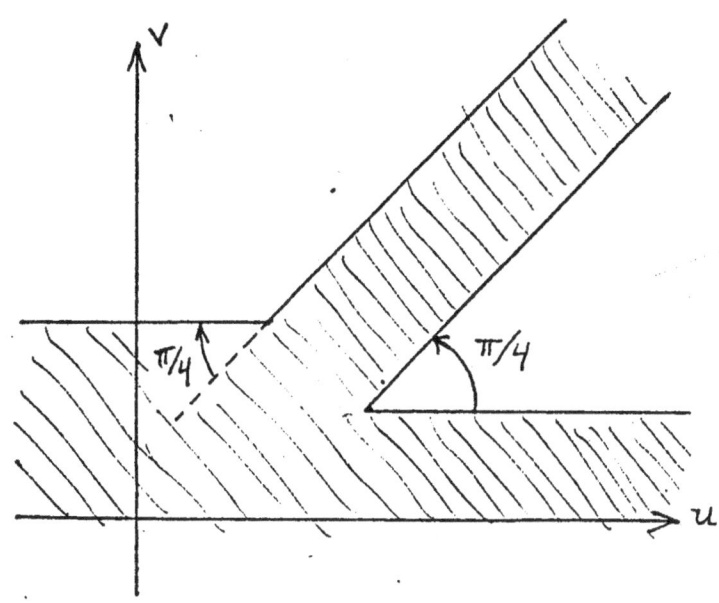

Solution

Now there are three vertices at infinity and two that are finite for a

total of five. If we select our starting point w_∞ on the real w-axis we can denote the vertices as shown:

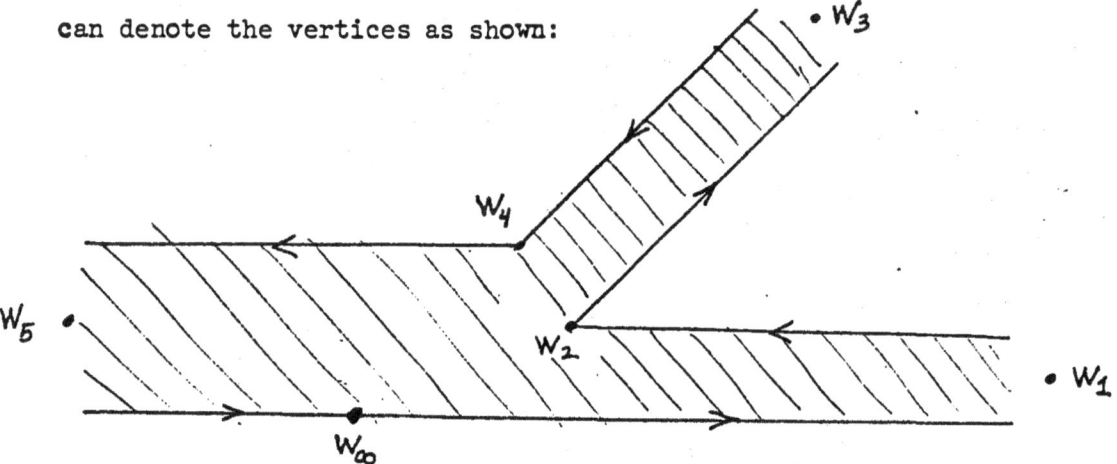

Of course, the vertices w_1, w_3 and w_5 are at infinity.

We must now determine the exterior angles at each vertex. The exterior

angle is defined as our change in direction as we pass through the vertex.

At w_1 then, we see that we must make a 180° turn in the counterclockwise

(positive) sense.

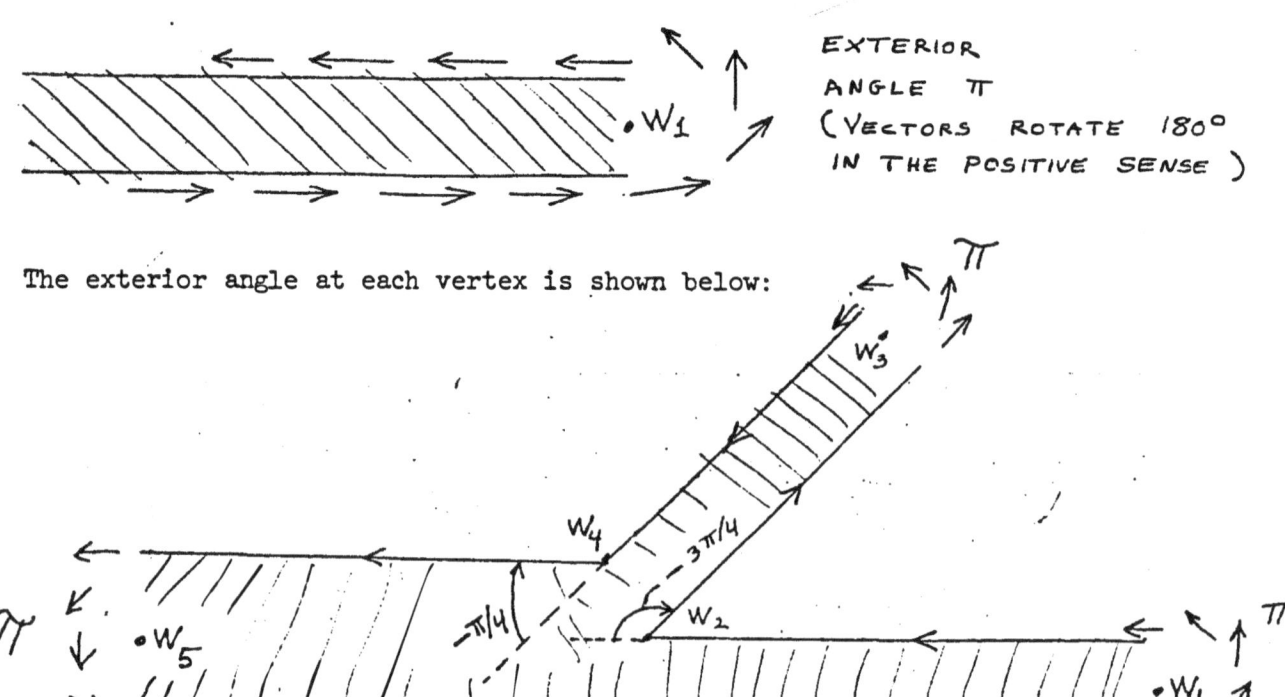

The exterior angle at each vertex is shown below:

Notice that the sum of the exterior angles

$$\pi - \frac{3\pi}{4} + \pi - \frac{\pi}{4} + \pi$$

equals 2π (as is always the case when the polygon is simply connected).

The following table summarizes our findings.

vertex w_i	exterior angle $\alpha_i \pi$	i
w_1	π	1
w_2	$-\frac{3\pi}{4}$	$-3/4$
w_3	π	1
w_4	$-\frac{\pi}{4}$	$-1/4$
w_5	π	1

We can now write the transformation (2) as

$$(4) \quad w = A \int^z \frac{(z - x_2)^{3/4}(z - x_4)^{1/4}}{(z - x_1)(z - x_3)(z - x_5)} \, dz + B$$

If we select $w = w_5$, then x_5 is at infinity and we see from (3) that the

factor $(z - x_5)$ is removed to get

$$w = A \int^z \frac{(z - x_2)^{3/4}(z - x_4)^{1/4}}{(z - x_1)(z - x_3)} \, dz + B.$$

Notice that any factor in the integrand of (4) would be deleted by selecting

the starting point w_∞ at the corresponding vertex on the polygon.

Problems:

Without specifying the constants A, B or the x_i's, write the two forms of the Schwarz-Christoffel transformation (2) and (3) for the mapping of the

upper half of the z-plane onto the polygon shown in the w-plane.

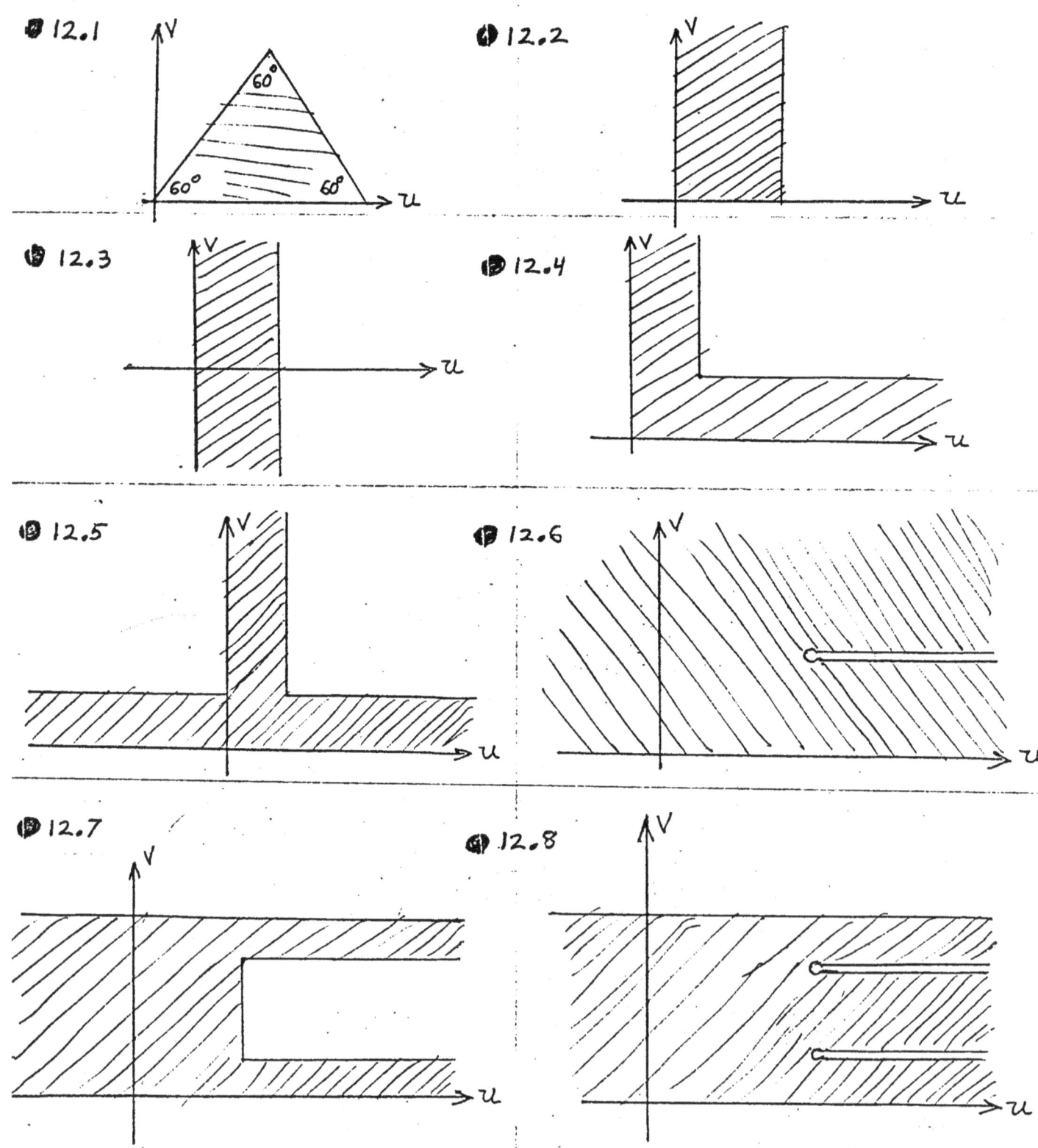

12.1   12.2   12.3   12.4   12.5   12.6   12.7   12.8

● 12.9 Suppose the points x_1, x_2, $\cdots$, x_n in the Schwarz-Christoffel formula

(2) are selected such that $x_1 > x_2 > \cdots > x_n$. What region in the z-plane

now maps onto the interior of the polygon in the w-plane?

In the previous examples, certain constants A, B and the x_i's of the

Schwarz-Christoffel formula were left undetermined. In the following examples

we illustrate how specific values for these constants can be selected.

Example 3

Use the Schwarz-Christoffel transformation to map $\text{Im}(z) \geq 0$ onto the

region shown.

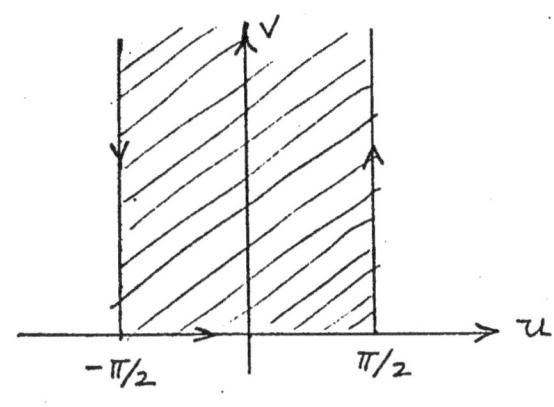

W- PLANE

Solution

This region has three

vertices $w_1 = -\pi/2$,

$w_2 = \pi/2$ and $w_3 = \infty$.

If we select our starting

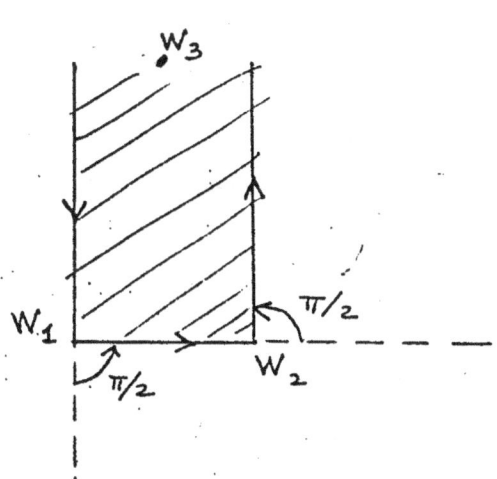

point at $w_\infty = w_3 = \infty$ we have from (3) with the exterior angles shown

$$w = A' \int^z (z - x_1)^{-1/2}(z - x_2)^{-1/2} \, dz + B.$$

Remark 4 above stated that any three of the x_i s are arbitrary. Since

we have only three vertices, each of x_1, x_2, x_3 can be selected at will.

We have already set $x_3 = \infty$. Suppose we select $x_1 = -1$ and $x_2 = 1$. We

now have

$$w = A' \int^z (z + 1)^{-1/2}(z - 1)^{1/2} \, dz + B.$$

Since $(z - 1)^{1/2} = (-1)^{1/2}(1 - z)^{1/2}$,

we can write the last expression as

$$w = A \int^z (1 + z)^{-1/2}(1 - z)^{-1/2} \, dz + B$$

$$W = A \int^z \frac{dz}{\sqrt{1-z^2}} + B$$

where we have simply changed A' to A because of the influence of the factor

$(-1)^{1/2}$.

Since

$$\int^z \frac{dz}{\sqrt{1 - z^2}} = \sin^{-1} z$$

we have

$$w = A \sin^{-1} z + B.$$

We must select A and B so that $w = -\pi/2$ when $z = -1$:

$$-\frac{\pi}{2} = A \sin^{-1}(-1) + B$$

$$-\frac{\pi}{2} = A(-\pi/2) + B.$$

We must have $w = \pi/2$ when $z = 1$:

$$\frac{\pi}{2} = A \sin^{-1}(-1) + B$$

$$\frac{\pi}{2} = \frac{A\pi}{2} + B.$$

We see that $A = 1$ and $B = 0$ satisfies these two equations and thus

$$w = \sin^{-1} z$$

is the appropriate mapping function.

Example 4

Map the region $\text{Im}(z) \geq 0$ onto the region shown.

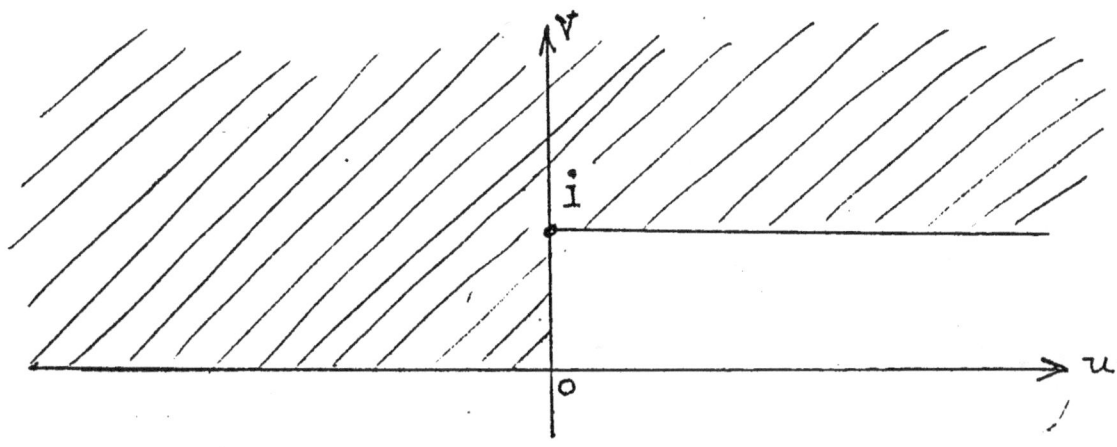

Solution

As in the previous example there are only three vertices w_1, w_2 and

w_3. If we select the

starting point at

$w_\infty = w_3$ we have

from (3)

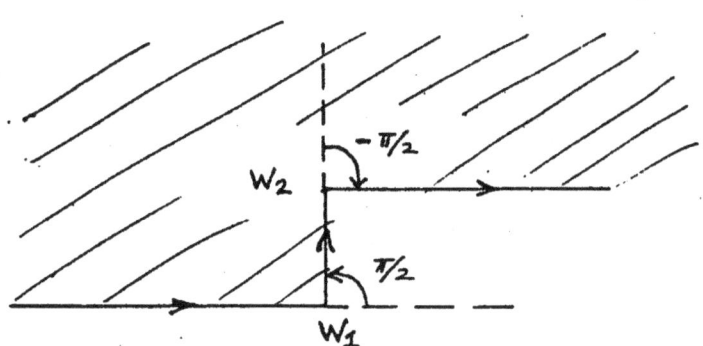

$$w = A' \int^z (z - x_1)^{-1/2} (z - x_2)^{1/2} \, dz$$

Now three of the x_i's are always arbitrary. (See Remark 4.) We have

selected $x_3 = \infty$, and now we take $x_1 = 0$ and $x_2 = 1$. We have

$$w = A' \int^z z^{-1/2} (z - 1)^{1/2} \, dz + B.$$

Writing

$$A' (z - 1)^{1/2} = A(1 - z)^{1/2}$$

we have

$$w = A \int^z \sqrt{\frac{1 - z}{z}} \, dz + B.$$

To evaluate this integral we set

$$z = \sin^2 \theta, \quad dz = 2 \sin \theta \cos \theta \, d\theta$$

and get

$$\int^z \sqrt{\frac{1 - z}{z}} \, dz = 2 \int^\theta \sqrt{\frac{1 - \sin^2 \theta}{\sin^2 \theta}} \, \sin \theta \cos \theta \, d\theta$$

$$= 2 \int_{\theta}^{\theta} \cos^2 \theta \, d\theta$$

$$= \int_{\theta}^{\theta} (1 + \cos 2\theta) \, d\theta$$

$$= \theta + \frac{\sin 2\theta}{2}$$

$$= \theta + \sin \theta \cos \theta$$

Since $\sin \theta = \sqrt{z}$ we have

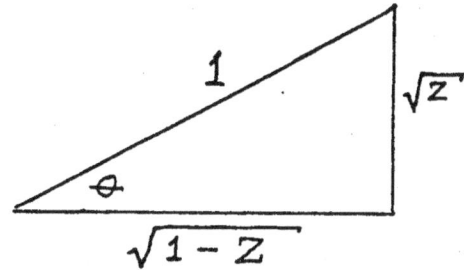

$$(5) \quad \int^{z} \sqrt{\frac{1-z}{z}} \, dz = \sin^{-1} \sqrt{z} + \sqrt{z} \sqrt{1-z} \, ,$$

and our mapping function is thus

$$w = A \left(\sin^{-1} \sqrt{z} + \sqrt{z(1-z)} \right) + B.$$

We previously made the correspondences

$$w_1 = 0 \longrightarrow x_1 = 0$$

and

$$w_2 = i \longrightarrow x_2 = 1.$$

The first of these yields $\quad 0 = A(\sin^{-1} 0 + 0) + B$ and therefore $B = 0$.

The second yields

$$i = A (\sin^{-1}(1) + 0)$$

$$i = \pi/2 \, A$$

and thus $A = 2i/\pi$. We have now our mapping function

$$(6) \quad w = \frac{2i}{\pi} (\sin^{-1} \sqrt{z} + \sqrt{z(1-z)}) .$$

Example 5

(a) Show that the point $z = i$ maps onto the point $w \approx -0.77 + 1.06\,i$

under the function (6) of the previous example.

(b) Picture the region shaded on the w-plane in the previous example as a

large ocean of fluid. The ocean floor is the step of height one shown. If

the fluid is moving with velocity V in a horizontal direction from left

to right far from the origin $w = 0$, determine the velocity of the fluid at

the point $w = -0.77 + 1.06\,i$

Solution

(a) We must first calculate (6) with $z = i$:

$$w = \frac{2i}{\pi} (\sin^{-1} \sqrt{i} + \sqrt{i(1-i)})$$

$$(7) \quad w = \frac{2i}{\pi} (\sin^{-1}(e^{i\,\pi/4}) + \sqrt{1+i})$$

The inverse sine can be approximated easily from the contour map of the sine

function in Figure 2.7. We see there that the contour lines

$\rho = 1$ and $\emptyset = \pi/4$ intersect at 0.57 + 0.75 i. Thus

$$\sin^{-1}(e^{i\,\pi/4}) \approx 0.57 + 0.75\,i$$

Now

$$\sqrt{1 + i} = \sqrt{\sqrt{2}\,e^{i5\pi/4}} = \sqrt[4]{2}\,(\cos 22.5^\circ + i \sin 22.5^\circ)$$

$$\approx 1.10 + .46\,i.$$

Simple calculation with (7) now yields

$$w \approx \frac{2i}{\pi}\ (.57 + .75\,i + 1.10 + .46\,i)$$

$$w \approx -.77 + 1.06\,i$$

(b)

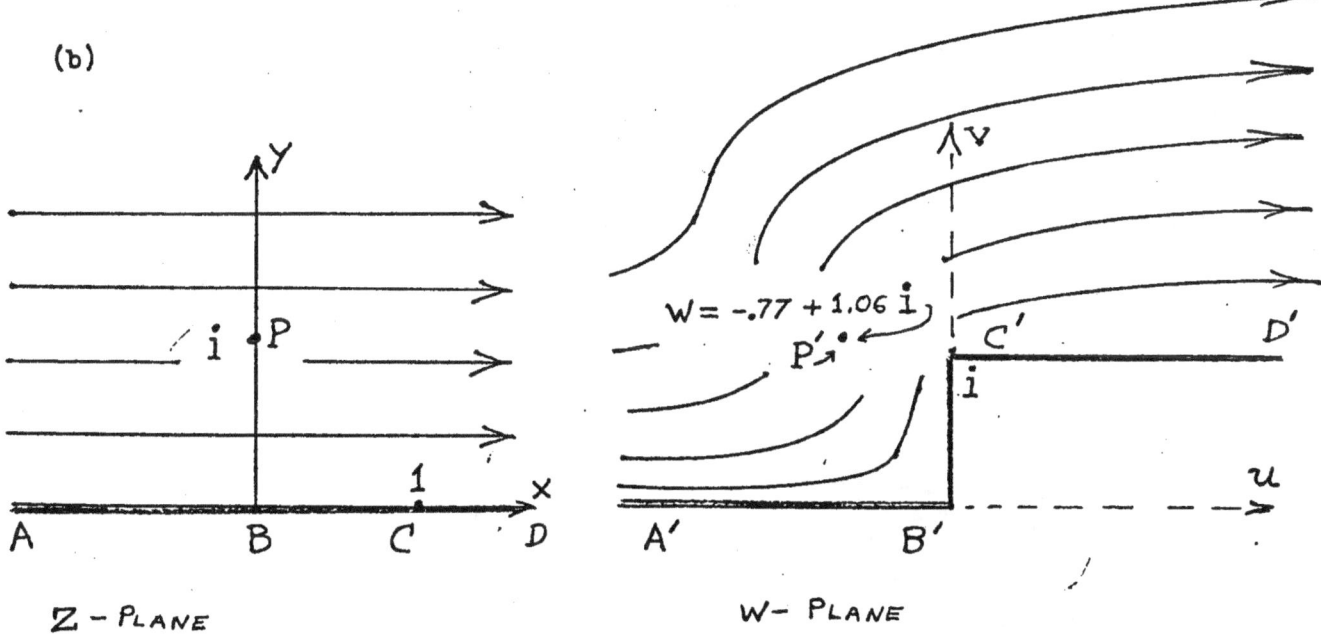

Z - PLANE   W- PLANE

The horizontal stream lines in the z-plane map onto the bent stream lines of the w-plane under the mapping function (6) of Example 4. The complex potential in the z-plane is then

$$f(z) = U \, z$$

where U is the velocity at infinity. If we could solve (6) for z in terms of w in the form $z = g(w)$ we could then write

$$f(g(w)) = U \, g(w)$$

as the complex potential in the w-plane. Unfortunately, there seems to be no easy way to write $g(w)$ explicitly. Nevertheless, we can still find the velocity, for the velocity in the w-plane is given by

$$\text{velocity} = \overline{\frac{df}{dw}} = U \, \overline{\frac{dg(w)}{dw}}$$

$$(8) \quad \text{velocity} = U \, \overline{\frac{dz}{dw}} \, .$$

We can compute $\frac{dw}{dz}$ from (6), but it is much easier to find this derivative from (5) as $\frac{dw}{dz} = \frac{2i}{\pi} \sqrt{\frac{1-z}{z}}$.

Therefore

$$\frac{dz}{dw} = \frac{\pi}{2i} \sqrt{\frac{z}{1-z}}$$

and (8) becomes

$$\text{velocity} = \frac{\pi i}{2} \sqrt{\frac{\bar{z}}{1 - \bar{z}}} \, U$$

Now we must select U so that when w is near ∞ , the velocity is V.

When w is near ∞ , z is also near ∞ and we have

$$\lim_{\bar{z} \to \infty} \sqrt{\frac{\bar{z}}{1 - \bar{z}}} = \sqrt{-1} = i.$$

Thus we have near infinity

$$\text{velocity} = \frac{\pi \cdot i}{2} (i) \; U = -\frac{\pi}{2} \; U$$

and we take $U = -\dfrac{2\,V}{\pi}$.

Thus we have

$$\text{velocity} = -Vi \sqrt{\frac{\bar{z}}{1 - \bar{z}}} \; .$$

Finally we find the velocity at $w \approx -.77 + 1.06\,i$ by substituting the

image of this point $(z = i)$ into this last expression to get

$$\begin{aligned}
\text{velocity} &= -Vi \sqrt{\frac{-i}{1 + i}} \\[4pt]
&= -Vi \sqrt{\frac{-1 - i}{2}} \\[4pt]
&= V \sqrt{\frac{1 + i}{2}} \\[4pt]
&\quad V (1.18 + .86\,i)
\end{aligned}$$

This vector has magnitude 1.46 V and makes angle 36° with the u-axis.

Example 6

Use the Schwarz-Christoffel transformation to map $\text{Im}(z) \geq 0$ onto the

region shown.

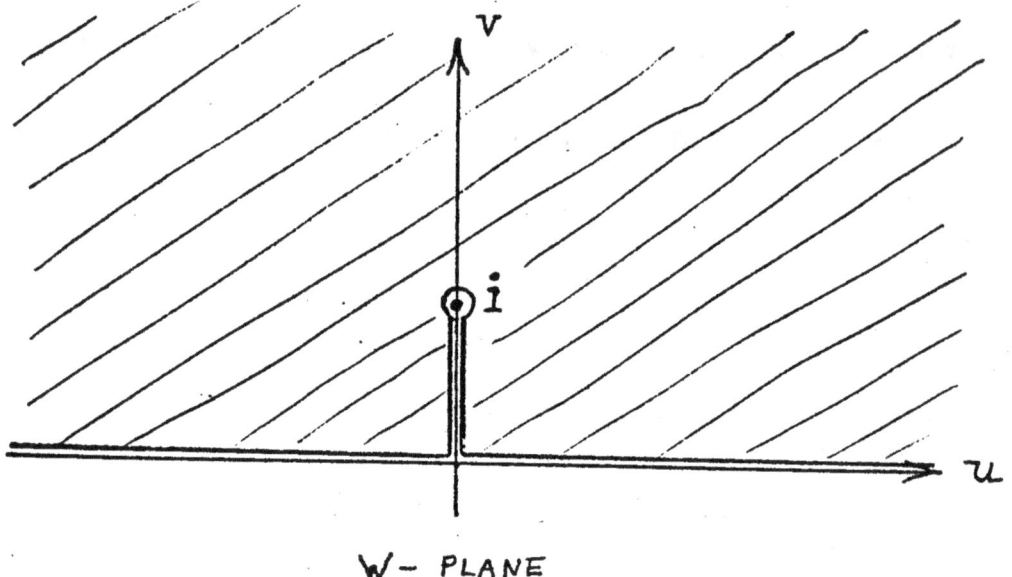

W — PLANE

Solution

This region in the w-plane has four vertices, $w_1 = 0$, $w_2 = i$, $w_3 = 0$

and $w_4 = \infty$. We are

always free to select

three of the points

x_1, x_2, x_3 and x_4

at will. Suppose we

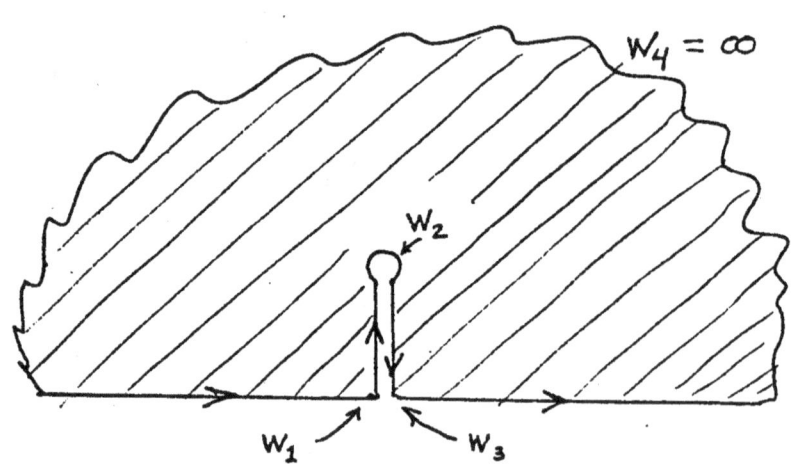

take $x_1 = -1$, $x_3 = 1$ and $x_4 = \infty$. We cannot arbitrarily specify the

number x_2 since it is a fourth value. All we know is that it is between

x_1 and x_3 and thus $-1 < x_2 < 1$.

The following table summarizes our selections.

vertices w_i	images x_i	exterior angles $\alpha_i \pi$	α_i
$w_1 = 0$	-1	$\pi/2$	$\frac{1}{2}$
$w_2 = i$	x_2	$-\pi$	-1
$w_3 = 0$	1	$\pi/2$	$\frac{1}{2}$
$w_4 = \infty$	∞	2π	2

Thus our mapping function is from (3)

$$w = A' \int^z \frac{(z - x_2)\, dz}{\sqrt{z^2 - 1}} + B$$

$$w = A \int^z \frac{z - x_2\, dz}{\sqrt{1 - z^2}} + B$$

$$w = A \int^z \frac{z\, dz}{\sqrt{1 - z^2}} - x_2 A \int^z \frac{dz}{\sqrt{1 - z^2}} + B$$

$$w = -A \sqrt{1 - z^2} - x_2\, A\, \sin^{-1} z + B.$$

We must now determine A, B and x_2. Since $w = 0$ when $z = -1$ we have

$$(9) \qquad 0 = \frac{\pi x_2\, A}{2} + B$$

Since $w = i$ when $z = x_2$ we have

$$(10) \qquad i = -A \sqrt{1 - x_2^2} - x_2\, A\, \sin^{-1} x_2 + B.$$

Since $w = 0$ when $z = 1$ we have

$$(11) \qquad 0 = -\frac{\pi x_2\, A}{2} + B.$$

Adding (10) and (11) we see that $B = 0$. With $B = 0$, (9) becomes $x_2 A = 0$.

This means that either $A = 0$ or $x_2 = 0$. If $A = 0$, the mapping becomes

the constant function $w = 0$ which is impossible. Thus $x_2 = 0$ and we

have from (10), $A = -i$. Thus we get

$$w = i \sqrt{1 - z^2} .$$

Problems

Use the Schwarz – Christoffel transformation to determine a function

which maps $\text{Im}(z) \geq 0$ onto the polygon shown in the w-plane.

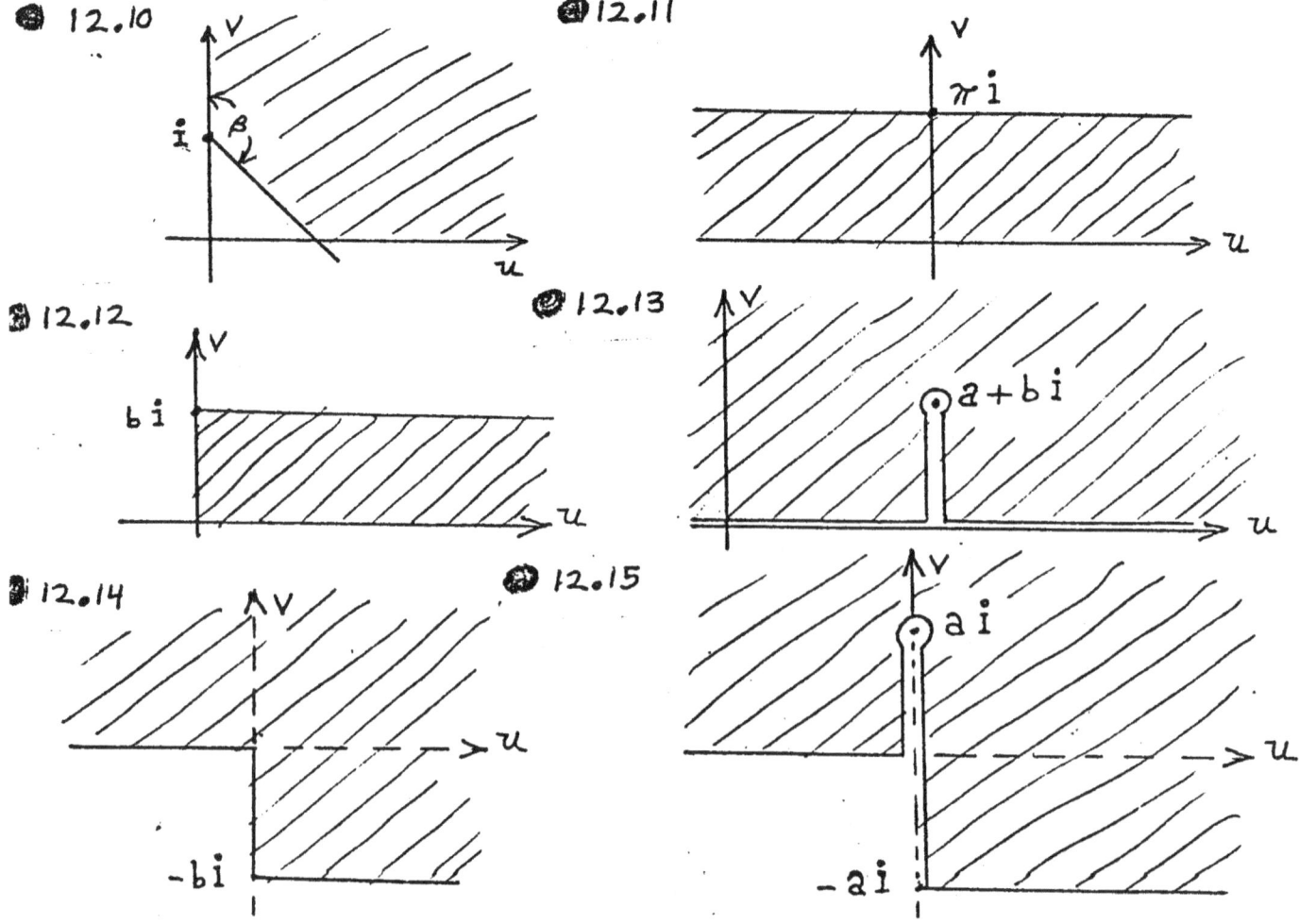

12.16   Determine the temperature

u(x, y) at each point of the

shaded region when the temperatures

on the boundary are as shown.

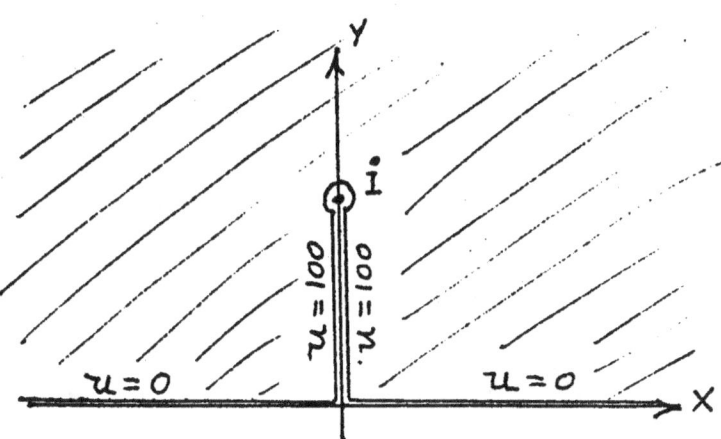

12.17   Determine the temperature

at the point (- 0.77, 1.06)

with the given boundary temperatures.

(See problem 46 and Examples 4 and 5.)

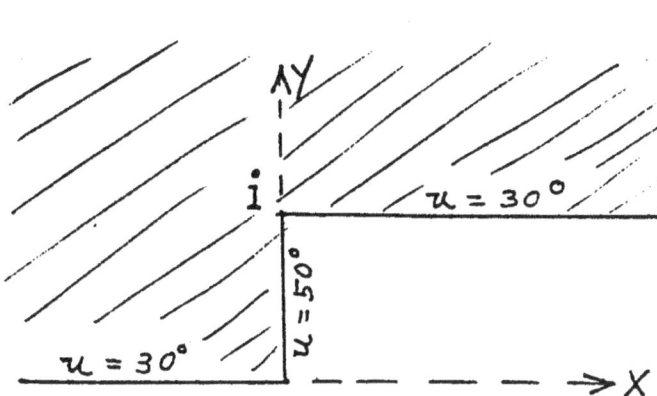

12.18   Describe the region on the w-plane which is the image of $\text{Im}(z) \geq 0$

under the mapping function

$$w = \int_0^z z^{-2/3}(1 - z)^{-2/3} \, dz.$$

APPENDIX I

SOLUTIONS TO PROBLEMS

Problems from Chapter 9

1/ Since $w = z = x + iy$, the stream lines are $y = c$ where c is any real number.

2/ Since $w = \log z = \log r + i\theta$

$$= \log \sqrt{x^2+y^2} + i \tan^{-1} \frac{y}{x},$$

the stream lines are the lines $\tan^{-1} \frac{y}{x} = c$, or $y = cx$, where c is a constant.

3/ Since $w = \frac{1}{z} = \frac{1}{x+iy} = \frac{x-iy}{x^2+y^2}$, we see that the stream lines are $\frac{-y}{x^2+y^2} = c$ or

$c(x^2+y^2) + y = 0$. Manipulating into the standard form for a circle we have

$$x^2 + y^2 + \frac{1}{c}y + \frac{1}{4c^2} = \frac{1}{4c^2}$$

$$x^2 + \left(y + \frac{1}{2c}\right)^2 = \left(\frac{1}{2c}\right)^2.$$

This is a circle with center at $\left(0, -\frac{1}{2c}\right)$ and passing through the origin.

4/ If we select $a = -1$ and $b = 1$ in Figure 9.1 (iv), we see that we have the appropriate flow.

Thus $\log \dfrac{z+1}{z-1} = \log(z+1) - \log(z-1) =$

$\log|z+1| - \log|z-1| + i\{\arg(z+1) - \arg(z-1)\}$.

The stream lines are then

$$\arg(z+1) - \arg(z-1) = C$$

$$\boxed{\tan^{-1}\dfrac{y}{x+1} - \tan^{-1}\dfrac{y}{x-1} = C}$$

5/ In Figure 9.1 (vii) we see that if we select $a = 1$ we have the appropriate stream lines. From Example 2 we know the the stream lines are $y - \dfrac{y}{x^2+y^2} = C$.

6/ The stagnation points occur where $f'(z) = 0$,

$$f(z) = Vz + m \log z$$
$$f'(z) = V + \dfrac{m}{z} = 0$$

Thus $z = -\dfrac{m}{V}$ is the stagnation point.

7/ (a) $f(z) = m z^2$, and thus the velocity is

$\overline{f'(z)} = 2m\bar{z} = 2m(x - iy)$. The stream lines are determined from

$$f(z) = m z^2 = m(x + iy)^2$$
$$= m(x^2 - y^2) + 2mxy\, i.$$

Thus $xy = c$ are the stream lines.

(b) Velocity $= \overline{w'} = \dfrac{m}{\bar{z}}$. The stream lines come from $w = m \log z = m \log r + im\theta$. Thus $\theta = c$ are the stream lines.

(c) Velocity $= \overline{w'} = \dfrac{-mi}{\bar{z}} = \dfrac{-mi\, z}{|z|^2}$

$= \dfrac{m(y + i x)}{x^2 + y^2}$. From

$$w = mi \log z = -m\theta + mi \log r$$

we see that $r = c$ is the equation of a stream lines.

(d) Velocity $= \overline{w'} = \dfrac{-m}{\bar{z} - \bar{b}} + \dfrac{m}{\bar{z} - \bar{a}}$.

Suppose we call $Re(a) = a_1$ and $Im(a) = a_2$, also $Re(b) = b_1$ and $Im(b) = b_2$. Then

$$w = m\log\left|\dfrac{z - a}{z - b}\right| + i\left[arg(z - a) - arg(z - b)\right] m$$

7/ (d) (continued)

$$w = m \log \left| \frac{z-a}{z-b} \right| = m i \left[\tan^{-1} \frac{y-a_2}{x-a_1} - \tan^{-1} \frac{y-b_2}{x-b_1} \right]$$

The stream lines are $\left[\tan^{-1} \frac{y-a_2}{x-a_1} - \tan^{-1} \frac{y-b_2}{x-b_1} \right] = c$,

(e) Velocity $= \overline{w'} = \dfrac{-m}{\bar{z}^2}$. The stream lines were determined in problem 3.

(f) Velocity $= \overline{w'} = V + \dfrac{m}{\bar{z}}$. Since

$$w = Vz + m \log z = Vx + m \log r + i(Vy + m\theta)$$

we see that $Vy + m \tan^{-1} \dfrac{y}{x} = c$ describes the stream lines.

8/ In Example 2 we saw that the velocity vector is

$$\overline{w'} = V\left(1 - \frac{a^2}{\bar{z}^2}\right) = V\left(1 - \frac{a^2}{r^2 e^{-i2\theta}}\right)$$

$$= V\left(1 - \frac{a^2 e^{i2\theta}}{r^2}\right)$$

$$= V\left[\left(1 - \frac{a^2}{r^2} \cos 2\theta\right) + i\frac{a^2}{r^2} \sin 2\theta\right].$$

Thus the square of the speed is

$$|\overline{w'}|^2 = V^2 \left[\left(1 - \frac{a^2}{r^2} \cos 2\theta\right)^2 + \frac{a^4}{r^4} \sin^2 2\theta\right]$$

8/ (continued)

$$|\overline{w'}|^2 = V^2\left[1 - \frac{2a^2\cos 2\theta}{r^2} + \frac{a^4}{r^4}\right]$$

We must maximize the above expression subject to the two conditions $r \geq a$ and $-\pi \leq \theta \leq \pi$. If we make $\cos 2\theta = -1$ then all three terms in brackets have a positive sign. This means $\theta = \pm\frac{\pi}{2}$. Since r is in the denominators, we let r assume its smallest value, a. Thus the points where the velocity is greatest are at $\pm ia$, the top and bottom of the cylinder. The velocity at these points is $2V$.

9/ The complex potential $2\log(z-i)$ describes a source at $(0,1)$ of the required strength while $-2\log z$ describes a sink at the origin of equal strength. Thus the complex potential is

$$f(z) = 2\log(z-i) - 2\log z$$
$$= 2\log\frac{z-i}{z} \quad ,$$

9/ (continued)

The velocity vector is

$$\overline{f'(z)} = \frac{2}{\overline{z}+i} - \frac{2}{\overline{z}} \quad,$$

Since

$$f(z) = 2 \log \left| \frac{z-i}{z} \right| + 2i \left[\tan^{-1} \frac{y-1}{x} - \tan^{-1} \frac{y}{x} \right]$$

$$= 2 \log \frac{\sqrt{x^2+(y-1)^2}}{\sqrt{x^2+y^2}} + 2i \left[\tan^{-1} \frac{y-1}{x} - \tan^{-1} \frac{y}{x} \right]$$

$$= \log \frac{x^2+(y-1)^2}{x^2+y^2} + 2i \left[\tan^{-1} \frac{y-1}{x} - \tan^{-1} \frac{y}{x} \right],$$

the equipotential lines are

$$\frac{x^2+(y-1)^2}{x^2+y^2} = c$$

$$x^2+y^2-2y+1 - cx^2 - cy^2 = 0$$

$$(1-c)(x^2+y^2) - 2y +1 = 0 \quad,$$

These are circles with center on the y-axis.
The stream lines are

$$\tan^{-1} \frac{y-1}{x} - \tan^{-1} \frac{y}{x} = c_1.$$

10/ From Example 5, we know that $\log(\cos z)$ is a complex potential with sources of strength $\frac{1}{2\pi}$ units of volume per unit time at the points where $\cos z = 0$. These points occur at $z = \frac{\pi}{2} + \pi n$, where $n = 0, \pm 1, \pm 2, \cdots$. The stagnation points occur where $f'(z) = 0$

$$f(z) = \log(\cos z)$$
$$f'(z) = \frac{-\sin z}{\cos z} = -\tan z .$$

Since $\tan z$ is zero for $z = n\pi$, where n is an integer, the stagnation points are at $z = n\pi$. The velocity vector is

$$\overline{f'(z)} = -\overline{\tan z} .$$

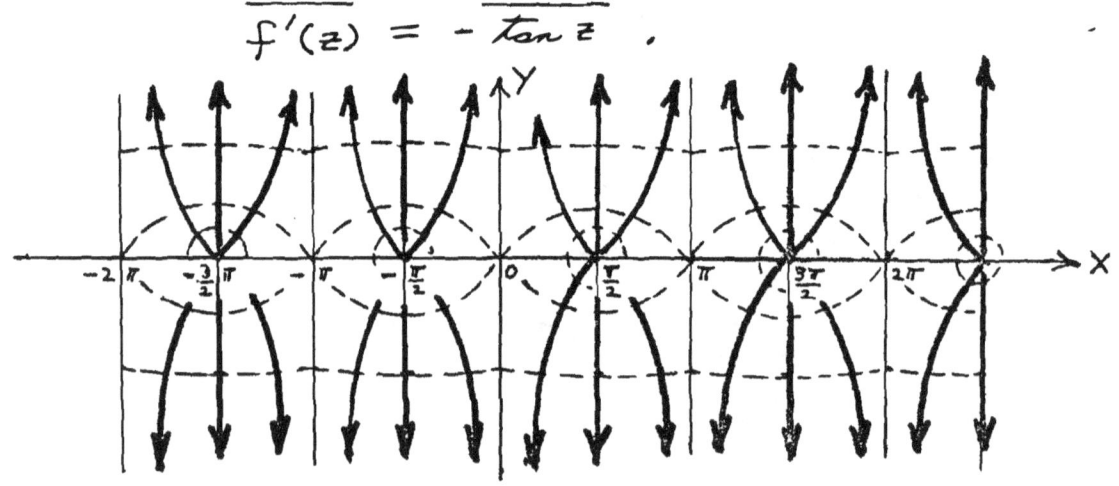

14/ We wish to take the complex potential

$$f(z) = 5\left(z + \frac{16}{z}\right)$$

and translate it upward 4 units. Thus $\mathfrak{Z} = M(z) = z + 4i$ and $z = \mathfrak{Z} - 4i$. Now we have

$$F(\mathfrak{Z}) = 5\left(\mathfrak{Z} - 4i + \frac{16}{\mathfrak{Z} - 4i}\right).$$

For large $|\mathfrak{Z}|$, $F(\mathfrak{Z}) \approx 5\mathfrak{Z}$ and no adjustment is necessary for the speed near infinity.

15/ We must take the complex potential

$$f(z) = 10\left(z + \frac{4}{z}\right)$$

and (i) rotate through the angle $90°$, and (ii) translate the origin to $4 + 4i$. Thus

$$\mathfrak{Z} = M(z) = e^{i\pi/2}(z + 4 + 4i) = i(z + 4 + 4i) \text{ and}$$

$z = -i\mathfrak{Z} - 4 - 4i$. The complex potential is now

$$F(\mathfrak{Z}) = 10\left(-i\mathfrak{Z} - 4 - 4i + \frac{4}{-i\mathfrak{Z} - 4 - 4i}\right).$$

For large $|\mathfrak{Z}|$, $F(\mathfrak{Z}) \approx -i10\mathfrak{Z}$, and $\overline{F'(\mathfrak{Z})} \approx 10i$ which is the required velocity.

12/ (continued)

volume per unit time. The velocity vector is

$$\overline{f'(z)} = -\overline{\frac{\cos z}{\sin z}} = -\overline{\cot z} \ , \quad \text{The stream}$$

lines look like those drawn for Example 5
except that their direction is reversed.

13/ The complex potential is

$$f(z) = \log(\tan \pi z) = \log\left(\frac{\sin \pi z}{\cos \pi z}\right)$$

$$= \log(\sin \pi z) - \log(\cos \pi z) \ .$$

Thus we have sources at $z = n$ and sinks
at $z = \frac{1}{2} + n$ of strength $\frac{1}{2\pi}$. The velocity
vector is

$$\overline{f'(z)} = +\overline{\frac{\pi \cos \pi z}{\sin \pi z}} - \overline{\frac{-\pi \sin \pi z}{\cos \pi z}} = \pi \left[\overline{\cot \pi z} + \overline{\tan \pi z}\right].$$

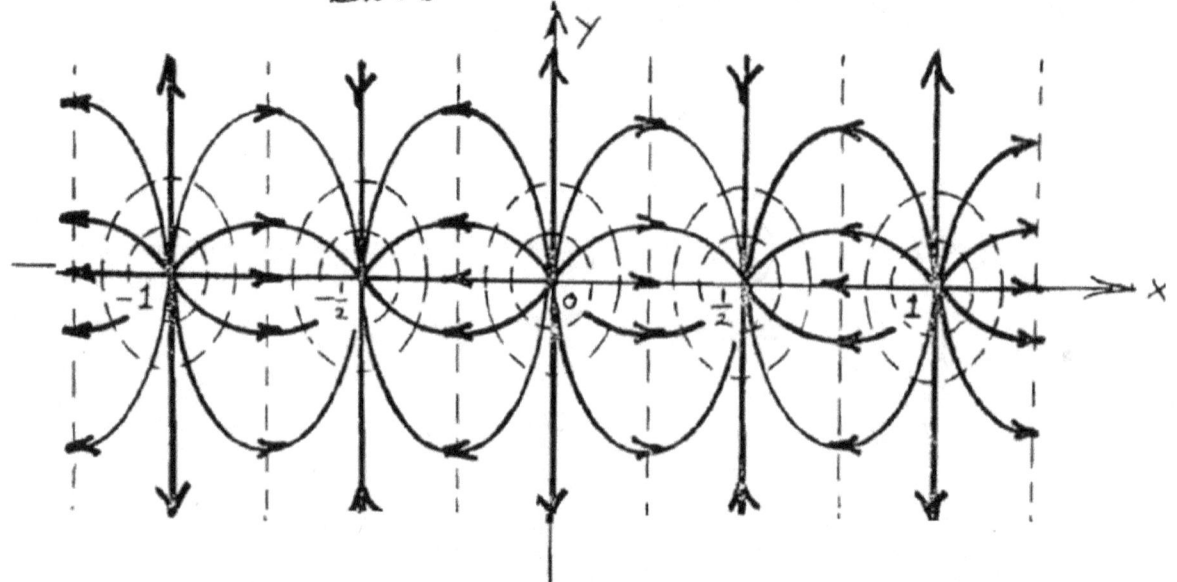

17/ (continued)

$$V_0\left(z+\tfrac{1}{z}\right) = V_0\left(e^{-i\pi/3}\,Z + \frac{e^{i\pi/3}}{Z}\right).$$

Using (5) of Example 2 we get the complex potential

$$F(\mathfrak{z}) = V_0\left(e^{-i\pi/3}\left(\mathfrak{z}+\sqrt{\mathfrak{z}^2-1}\right) + \frac{e^{i\pi/3}}{\mathfrak{z}+\sqrt{\mathfrak{z}^2-1}}\right).$$

The stagnation points $z = \pm 1$ in the z-plane map onto $Z = \pm e^{i\pi/3}$ in the Z-plane. These in turn map onto

$$\mathfrak{z} = \tfrac{1}{2}\left(\pm e^{i\pi/3} \pm e^{-i\pi/3}\right)$$

$$= \pm\tfrac{1}{2}\left(e^{i\pi/3} + e^{-i\pi/3}\right) = \pm \cos\tfrac{\pi}{3}$$

$$= \pm\tfrac{1}{2}.$$

Thus there are stagnation points at $\mathfrak{z} \pm \tfrac{1}{2}$ in the new flow. As in Example 2, the constant V_0 is selected as $V_0 = \frac{V}{2}$.

18/ This problem is similar to Example 3 except that now $Z = z$. Thus the complex potential is

16/ On the z-plane, the circle ABDF is described by $z = ce^{i\theta}$, where $1 < C$. The mapping of this circle is

$$s = \frac{1}{2}\left(z + \frac{1}{z}\right) = \frac{1}{2}\left(ce^{i\theta} + \frac{e^{-i\theta}}{c}\right)$$

$$= \frac{1}{2}\left(c(\cos\theta + i\sin\theta) + \frac{1}{c}(\cos\theta - i\sin\theta)\right)$$

$$= \frac{1}{2}\left(\left(c + \frac{1}{c}\right)\cos\theta + \left(c - \frac{1}{c}\right)\sin\theta \; i\right)$$

Thus
$$s = \frac{1}{2}\left(c + \frac{1}{c}\right)\cos\theta$$

$$t = \frac{1}{2}\left(c - \frac{1}{c}\right)\sin\theta$$

and these are the equations of the ellipse described in the figure. In Example 1, it was demonstrated that the mapping is conformal for $|z| \geq c$.

17/ This problem is similar to Example 2 with the exception that the mapping from the z-plane to the Z-plane is given by $Z = e^{i\pi/3}z$ and thus $z = e^{-i\pi/3}Z$. The complex potential now becomes

19/ (continued)

the desired flow. The desired mapping
is thus $\mathcal{S} = e^{-i\frac{\pi}{2}} b z = -i b z$ and

$z = \dfrac{i \mathcal{S}}{b}$. The complex potential is now

$$f(z) = F(\mathcal{S}) = -\frac{iV}{2}\left[\frac{i\mathcal{S}}{b} + \sqrt{-\frac{\mathcal{S}^2}{b^2} - 1} - \frac{1}{\frac{i\mathcal{S}}{b} + \sqrt{-\frac{\mathcal{S}^2}{b^2} - 1}} \right]$$

$$= \frac{V}{2b}\left[\mathcal{S} + \sqrt{\mathcal{S}^2 + b^2} - \frac{b^2}{\mathcal{S} + \sqrt{\mathcal{S}^2 + b^2}} \right]$$

9.2.1   Let $u(x, y, t)$ denote the temperature at the point (x, y) at

time t . Interpret each term in the "heat equation"

$$k \, \nabla^2 u = \frac{\partial u}{\partial t}$$

where k is a positive constant reflecting the thermal properties of the

substance. Review the discussion of the one dimensional heat equation in

section 8.6 .

9.2.2   Let $u(x, y, t)$ denote the vertical dispacement of the skin of a

drum head at point (x, y) and at time t . Interpret each term in the

"wave equation"

$$k^2 \, \nabla^2 u = \frac{\partial^2 u}{\partial t^2}$$

where k^2 is a constant reflecting the density and tension in the skin of

the drum head. Why is the wave equation an application of Newton's force

equals mass times acceleration law. Review section 8.7 in which the one

dimensional wave equation is examined.

9.2.3   Using Taylor's theorem in three variables

$$u(x+h, y+k, z+1) =$$

$$\sum_{m=0}^{\infty} \sum_{h=0}^{\infty} \sum_{p=0}^{\infty} \frac{\partial^{m+n+p} u(x,y,z)}{\partial x^m \, \partial y^n \, \partial z^p} \frac{h^m \, k^n \, 1^p}{m! \; n! \; p!}$$

show that the three dimensional Laplacian

$$\nabla^2 u = u_{xx} + u_{yy} + u_{zz}$$

measures the difference between u at the point (x, y, z) and the average

of u in an infinitesimal neighborhood of the point.

9.3.1   Find the stream functions for the flows described in Figure 9.1

diagrams (i v) , (v) and (viii) .

Answers   (i v) $v = - \tan^{-1} (y/x)$;   (v) $v = \tan^{-1} \left(\dfrac{y - a_2}{x - a_1} \right)$

$- \tan^{-1} \left(\dfrac{y - b_2}{x - b_1} \right)$ where $a = a_1 + i\, a_2$, $b = b_1 + i\, b_2$;

(vi) $v = V y + m \; \tan^{-1} \left(\dfrac{y}{x} \right)$.

9.3.2   Find the stream functions, sketch the stream lines, and locate

stagnation points for fluid motion described by:

(a) $z^2 - 2z$, (b) z^3 , (c) e^z .

9.4.1   Find the velocity vector for each of the flows in problem 9.3.2 .

9.4.2   Fluid enters the region shown in Example 3 of section 9.3 at the

rate of 8 cubic feet per second. Determine (a) the complex potential

and (b) the velocity vector.

Answers: (a) $\dfrac{16}{\pi} \log z$,   (b) $\dfrac{16}{\pi \bar{z}}$.

9.4.3   Fluid emerges from the point (2, 0) at the rate of 8π ft^3/sec. and

is absorbed at (0, 2) at the rate of 4π ft^3/sec.

(a) Find the complex potential.

(b) Find the velocity vector.

(c) Find the equations of the stream lines and the equipotential lines.

9.4.4   Discuss the flow described by each of the following complex potentials.

Locate all sources and sinks. Find all stagnation points. Find the velocity

vector and sketch the stream lines and the equipotential lines.

(a) $w = \log (\sin \pi z)$

(b) $w = \log (\cosh z)$

(c) $w = \log (\cot z)$

(d) $w = \log (\sec z)$

9.4.5   Consider the flow described by the complex potential

$$w = V \left(z + \frac{a^2}{z} \right) + m\,i\,\log z$$

(a) Show that we can consider the cylinder $|z| = a$ to be an obstacle in the fluid.

(b) Find the stagnation points and show that there are three cases:

(i) $m < 2\,a\,V$ in which case there are two stagnation points on the cylinder;

(ii) $m = 2\,a\,V$ and there is a stagnation point at $z = -a\,i$;

(iii) $m > 2\,a\,V$ and the stagnation point is below the cylinder.

(c) Find the equations of the stream lines and the equipotential lines.

(d) Find the velocity vector.

(e) Sketch the stream lines.

Answers:

(c) $v = V \left(r - \dfrac{a^2}{r} \right) \sin \theta + m \log r = c_1 \cdot,$

$u = V \left(r + \dfrac{a^2}{r} \right) \cos \theta - m\,\theta = c_2$

(d) $V \left(1 - \dfrac{a^2}{z^2} \right) + \dfrac{m\,i}{z}$

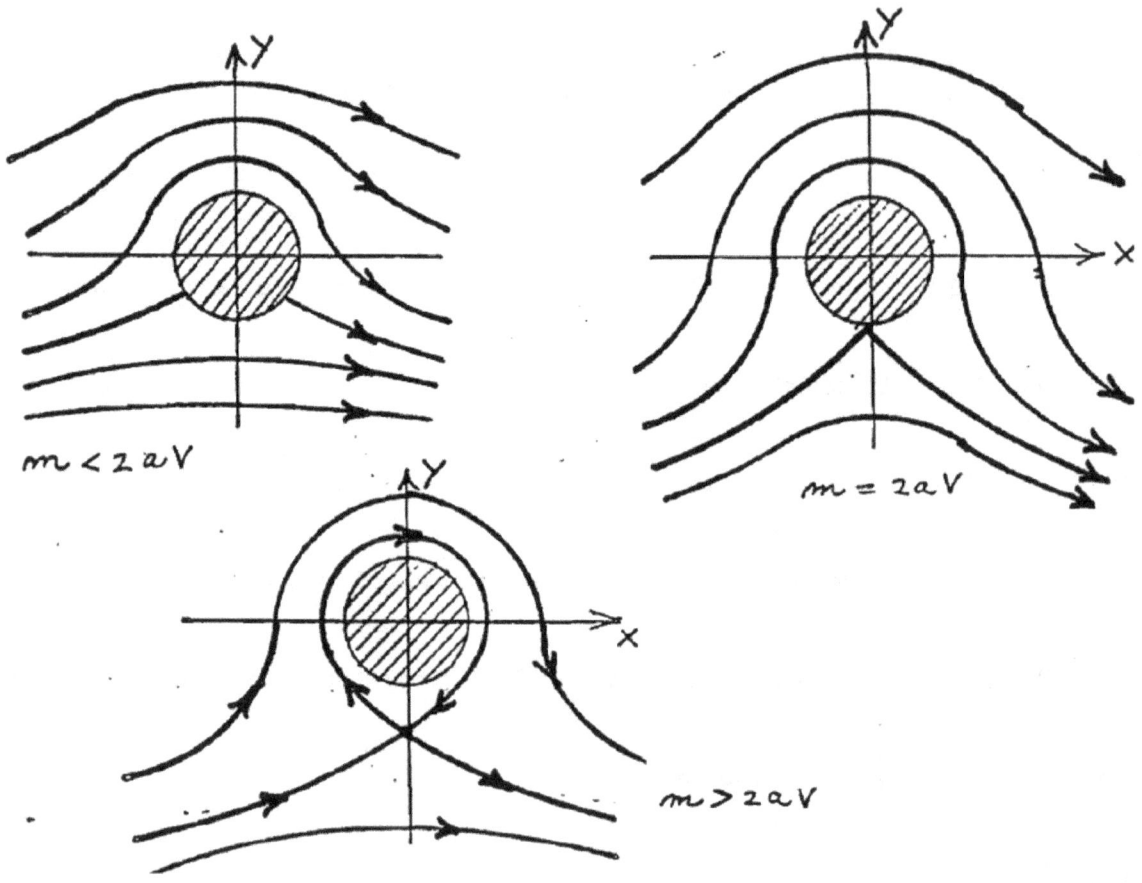

$m < 2aV$

$m = 2aV$

$m > 2aV$

9.4.6   A source of strength $\frac{m}{\epsilon}$ is located at $(\epsilon, 0)$ and a sink of

strength $\frac{m}{\epsilon}$ is located at $(-\epsilon, 0)$.

(a) Determine the complex potential.

$$\text{answer} \qquad \frac{m}{\epsilon} \qquad \log \quad \frac{z - \epsilon}{z + \epsilon} \quad .$$

(b) Show that as the source and the sink near each other, and their strengths

increase $(\epsilon \rightarrow 0)$, the complex potential just found approaches $\dfrac{-2m}{z}$

which is a dipole at the origin. (See Figure 9.1 (vi)).

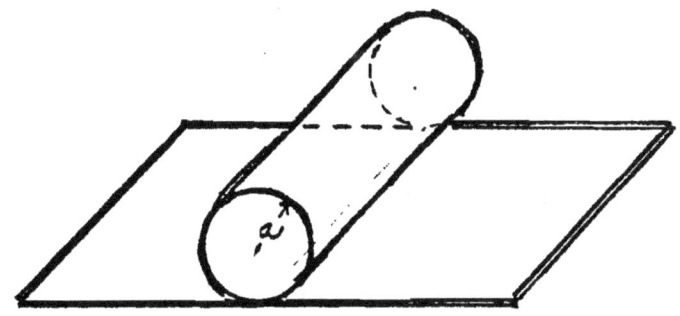

9.4.7 A log of radius "a" is placed

at the bottom of a deep stream bed in

which water flows from left to right

with uniform velocity V. In this

problem we will show that the complex

potential describing this flow is

$$f(z) = \pi a \, V \coth\left(\frac{\pi a}{z}\right).$$

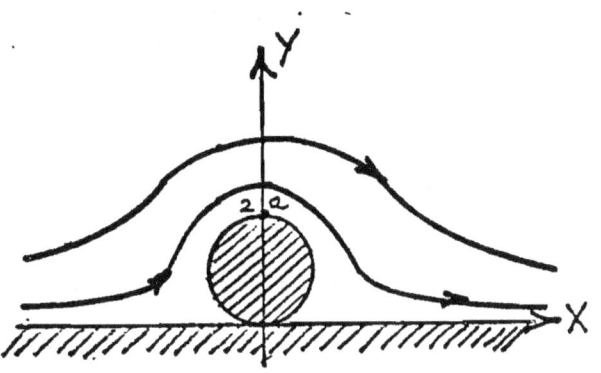

(a) Show that if $f = u + i v$, then the stream line $v = 0$ is the stream

bed ($y = 0$) and the log ($|z - i a| = a$).

(b) Show that for large z , $f'(z) \approx V$. This means that far from the

origin, the velocity is uniform from left to right with speed V.

9.5.1   A cylindrical obstacle of radius one is placed in a uniform flow of velocity V making the angle α with the real axis. Show that the complex potential is $V \left(z\, e^{-i\alpha} + \dfrac{e^{i\alpha}}{z} \right)$.

9.5.2 A cylindrical obstacle of radius 3 has its center placed at the point (3, - 3) of a uniform flow with velocity $8\, e^{i\,\pi/3}$. Find the complex potential describing the flow.

9.6.1   A flat plate of width 2b is placed in a uniform flow with velocity. V i. Find the complex potential describing the fluid flow.

9.6.2   A uniform flow with velocity V i encounters the elliptical obstacle of Example 3, section 9.6. Find the complex potential describing the flow.

9.6.3   It is possible to construct the mapping effected by the Joukowski transformation geometrically. Suppose z is located at the point P shown in the figure. Draw the line segment O P and measure its length r and angle θ . From the point P draw the line segment PQ making angle $-\theta$ as shown. Calculate 1/r and lay off this distance from P along PQ to

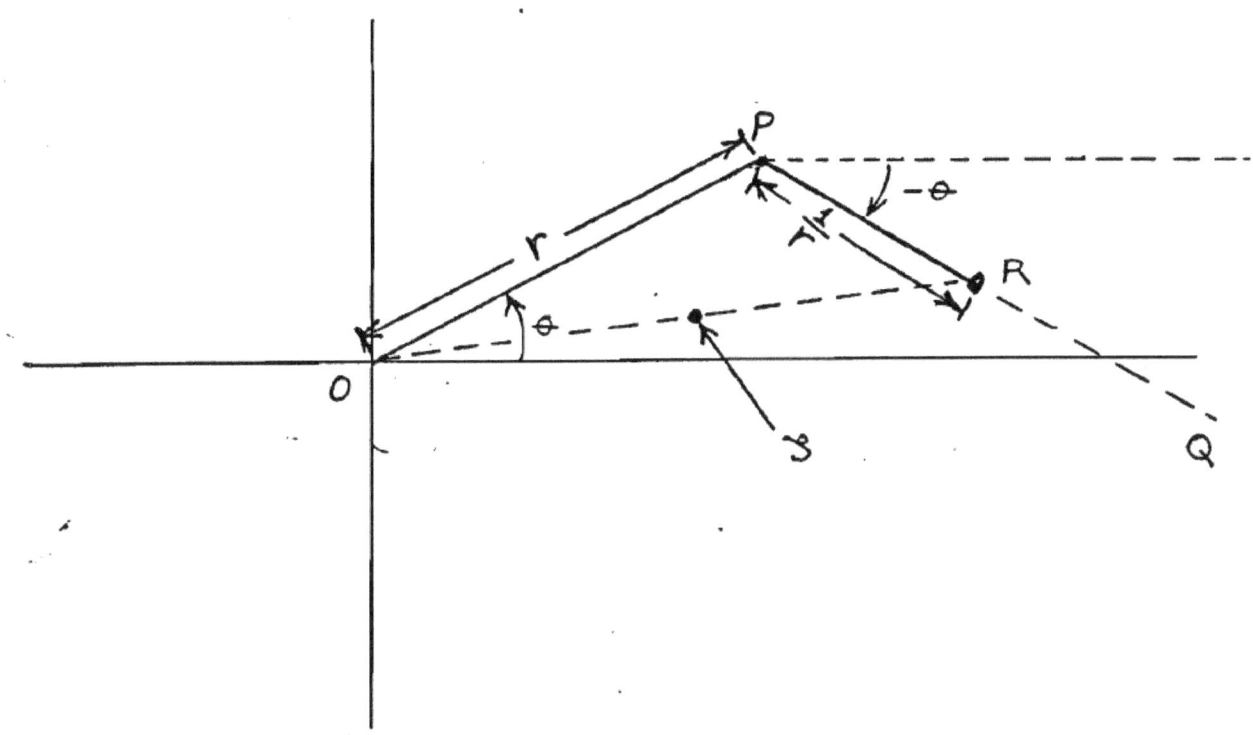

locate the point R. This point R is at $z + \frac{1}{z}$. Now draw the line

segment O R and bisect it to determine the point $\zeta = \frac{1}{2}\left(z + \frac{1}{z}\right)$.

Demonstrate the mappings shown in Figures 9.2 (iv), (v) and (vi), by

drawing the appropriate circles on the z - plane. After selecting various

points on these circles, determine their mapping by the above construction

and obtain the profiles in the ζ - plane.

9.6.4   A cylindrical obstacle whose cross-sectional area is the Joukowski

profile shown in Figure 9.2 (vi) is placed in a uniform stream with

velocity $V\, e^{i\alpha}$. Find the (a) complex potential and determine the (b)

stagnation points.

Answers: (a) $\dfrac{V}{2}\left[z\, e^{-i\alpha} + \dfrac{c^2 e^{i\alpha}}{z}\right]$ where

$c = \sqrt{(1+a)^2 + b^2}$ and $z = \sqrt{\mathcal{S} + \sqrt{\mathcal{S}^2 - 1}} + a - i\,b.$

(b) $\mathcal{S} = \tfrac{1}{2}\left[\pm\, c\, e^{i\alpha} - a + i\,b + \dfrac{1}{\pm\, c\, e^{i\alpha} - a + i\,b}\right].$

9.7.1   Find the potential u(x, y) and the complex potential f(z) =

u(x, y) + i v(x, y) in the upper half plane when conducting plates,

separated by insulation on the x-axis are charged to the potentials shown.

(a) ———— $u = -2$ ————○———— $u = 6$ ————→ x

(b) $u = 0$ ————○———— $u = V$ ————○———— $u = 0$ ————→ x
                 -1 1

(c) $u = 3$ ——○—— $u = 2$ ┃ $u = 1$ ○—— $u = 0$ ————→ x
               -5 0 5

9.7.2   Find the potential in the

first quadrant when conducting plates

on the positive real and imaginary

$\mathfrak{Z}$ - axies are charged as shown.

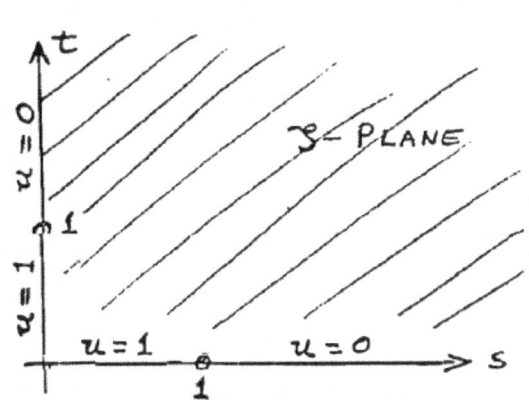

Hint: Map the solution of problem 9.7.1 (b) onto this region.

In problems 9.7.3 through 9.7.8 find (a) the complex potential, (b)

the electrostatic potential, (c) the lines of force and (d) the electric

intensity vector for each of the given distributions of line charges.

9.7.3   A line of charge q per unit length is at z = a and a line of

charge - q per unit length is at z = - a.

9.7.4   Identical lines of charge q per unit length are located at

i(π/2 + π n) where n is any integer.

9.7.5   Identical lines of charge q per unit length are located at

z = n/2 where n is any integer.

9.7.6   Lines of charge q are located at $\pi/2 + \pi n$ while lines of

charge -q are located at πn where n is any integer.

9.7.7   Lines of charge q are located at the points z = a + b n, where

a and b are real numbers and n is any integer.

9.7.8   Lines of charge q per unit length are located at the points z =

πn + i, where n is any integer, while the x-axis is a grounded con-

ducting plate.

9.7.9   The diagram shows two

lines of charge q per unit

length at $\Im = \pi/3$ i and

$\Im = 2\pi/3$ t. Both boundaries

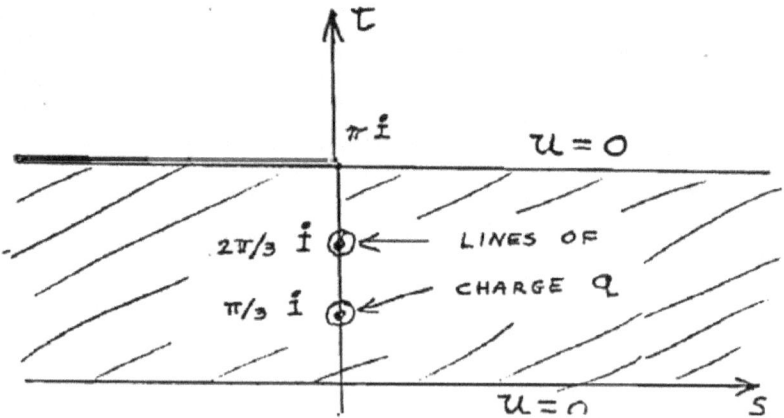

of the channel are grounded con-

ducting plates. Find the potential in the channel.

9.7.10   The diagram shows a line of

charge q per unit length located at

$\Im = a\ e^{iB}$ between two infinite

grounded conducting plates

making angle α. Find the

potential.

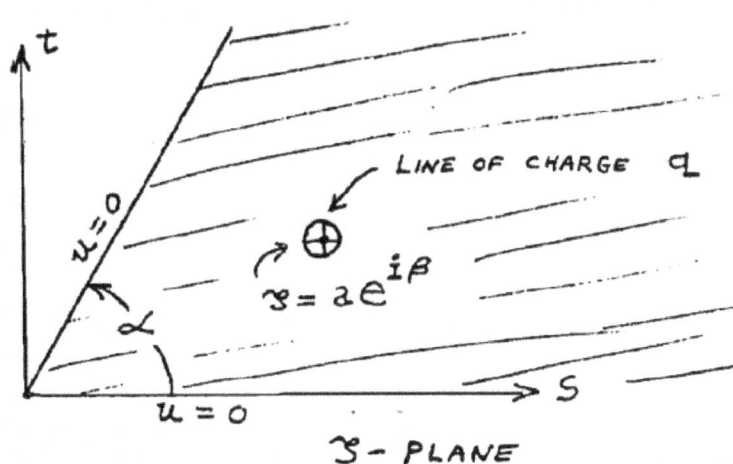

LINE OF CHARGE q

$s = a e^{i\beta}$

$u = 0$

$u = 0$

s - PLANE

9.7.11   (a) Find the potential

in the infinite channel shown when

the lower boundary is a grounded conducting plate and the upper boundary

is a conducting plate at potential $u = 1$.

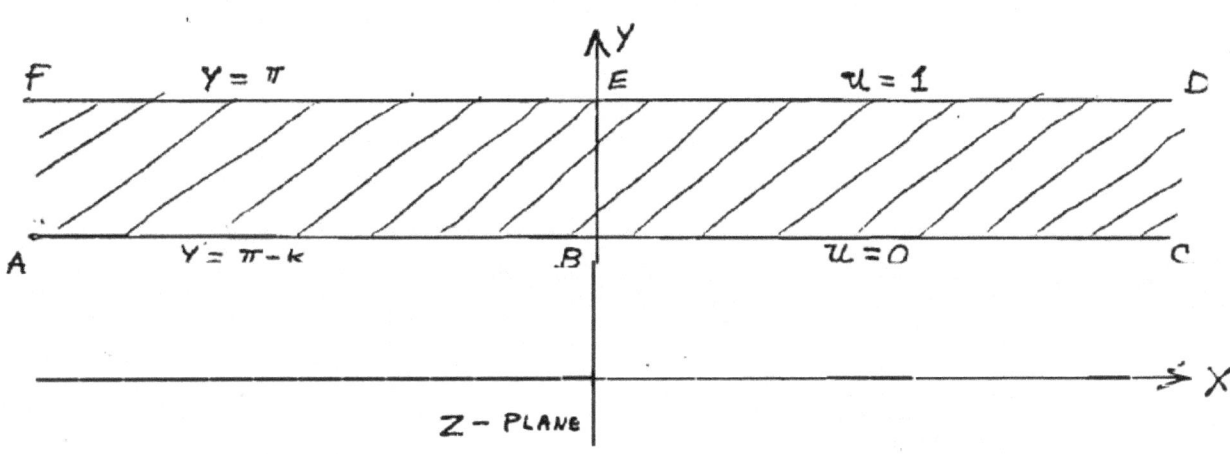

F $Y = \pi$ E $u = 1$ D

A $Y = \pi - k$ B $u = 0$ C

Z - PLANE

(b) Show that the mapping function $z = \log \dfrac{s - 1}{s + 1}$ maps the above channel

onto the lens shaped region

shown in the s - plane.

(c) Find the potential within

this lens shaped region when

the circular arc $A' B' C'$ is a

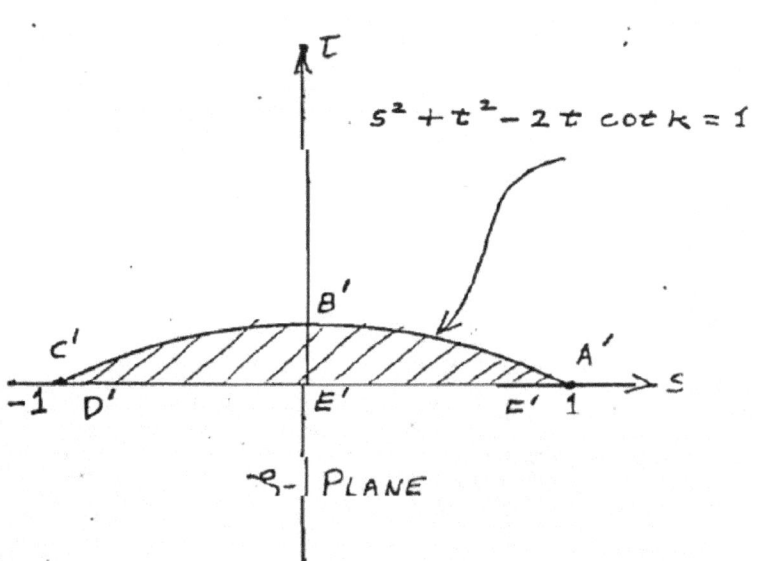

$s^2 + t^2 - 2t \cot k = 1$

B'

C' A'

-1 D' E' E' 1

s

s - PLANE

grounded conducting plate while the lower conducting plate $D'E'F'$ is

maintained at potential $u = 1$.

Find the bounded temperature inside each of the shaded regions shown

in problems 9.8.1 to 9.8.8. Describe the isotherms and the lines of flux.

9.8.1

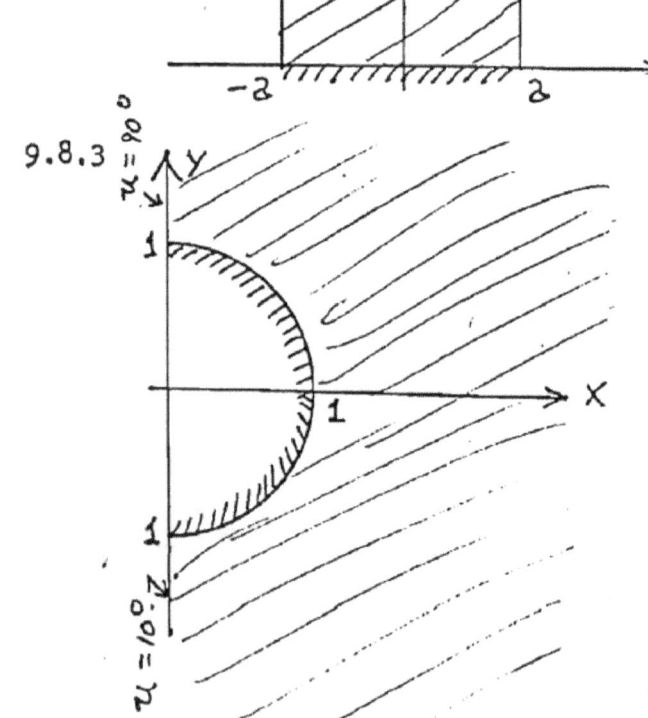

ANSWER:

$$u = \frac{40x}{2} + 50$$

9..8.2

9.8.3

ANSWER:

$$u = \frac{80}{\pi} \tan^{-1} \frac{y}{x} + 50.$$

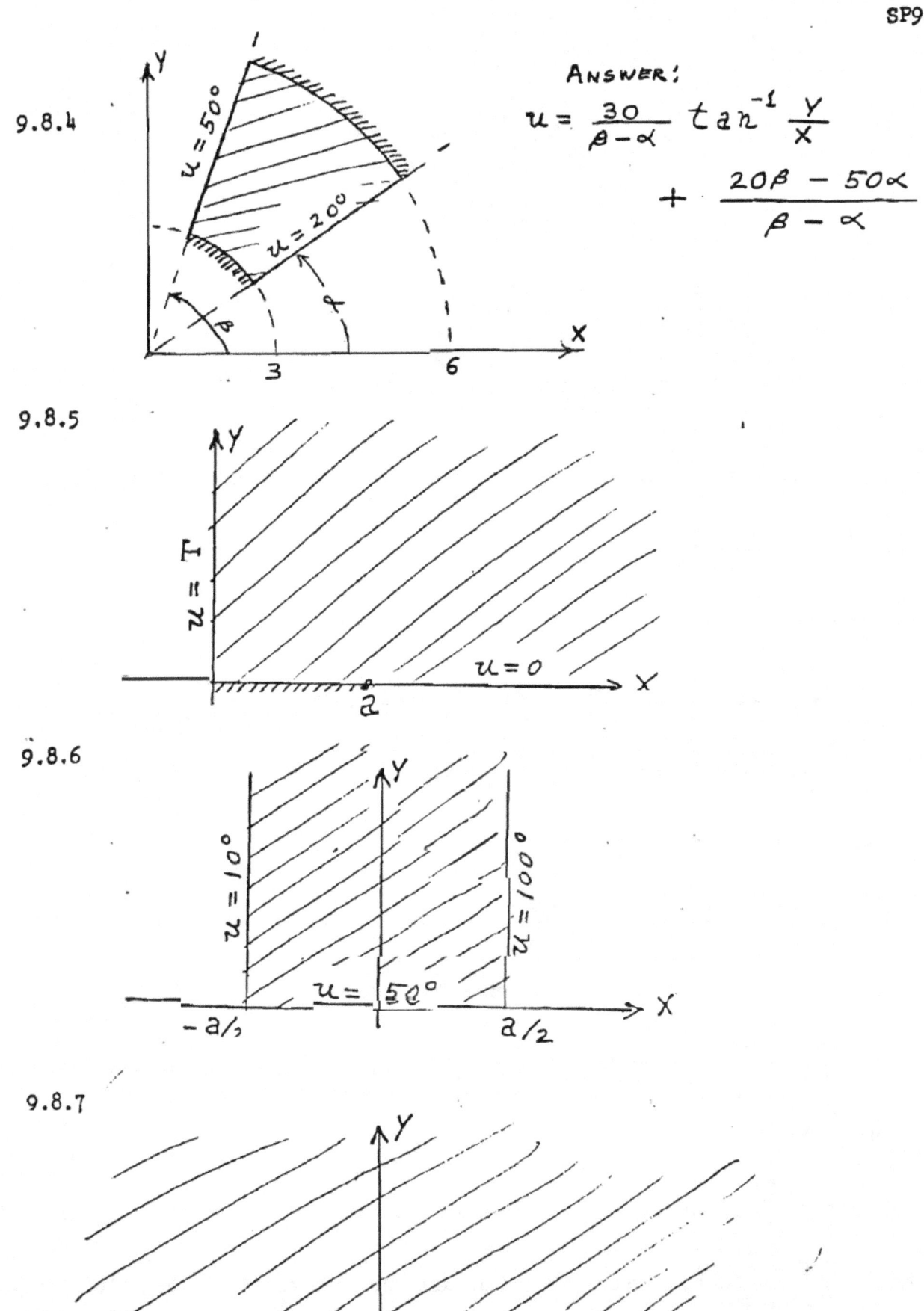

9.8.4

$u = 50°$

$u = 20°$

3

6

ANSWER:

$$u = \frac{30}{\beta - \alpha} \tan^{-1} \frac{Y}{X}$$

$$+ \ \frac{20\beta - 50\alpha}{\beta - \alpha}$$

9.8.5

$u = T$

$u = 0$

a

9.8.6

$u = 10°$

$u = 100°$

$u = 50°$

$-a/2$

$a/2$

9.8.7

$-b$

b

$u = 10°$

$u = 90°$

9.8.8

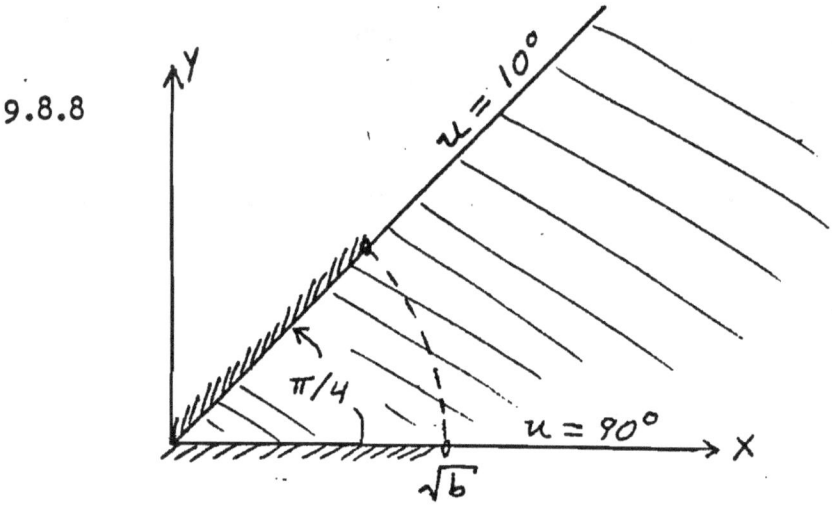

9.8.9   In this problem we examine the connection between relations (1)

through (7) of section 9.8.

(a) Derive relations (2) and (3) from (1).

(b) Derive relations (6) and (7) from (2) and (3).

(c) Show that the foci of the ellipses and hyperbolas described by (6) and

(7) are at the points ($\pm$ 1, 0).

(d) Derive (4). The simplest way to see this is to use the defining

relation for a hyperbola in the form: "The locus of all points P such

that $\overline{F_1 P} - \overline{F_2 P}$ = d, where F_1 and F_2 are the foci and d is the distance

between the vertic es."

(e) Derive (5). Note that an ellipse is the locus of all points P such

that $\overline{F_1 P} + \overline{F_2 P}$ = m, where F_1 and F_2 are the foci and m is the major axis.

9.9.1   Find the mapping of each of the following regions under the inversion transformation $w = 1/z$.

(a) $|z + 2| \leq 1$,　　　　(b) $|z + 3 i| \leq 1$,

(c) $|z + 2| \leq 2$,　　　　(d) $|z + \sqrt{3} - i| \geq 2$,

(e) $|z - 1 + i| \geq 5\sqrt{2}$,　(f)　$\text{Im}(z) \geq 1$,

(g)　$\text{Re}(z) \leq 0$.

9.9.2　　Use the transformation

$$w = -i \left(\frac{z - 1}{z + 1} \right)$$

to map the regions　(a)　$|z| \leq 4$　and　(b)　$|z - i| \leq 1$.

9.9.3　　Map the region　$\text{Im}(z) \geq 1$ under the transformation　$w = \frac{z - 3}{z + 2}$.

9.9.4　　Map the first quadrant in the　z-plane under the transformation

$$w = \frac{z - i}{z + i} .$$

Answer:　The lower half of the circle $|w| \leq 1$.

In problems　9.9.5　through　9.9.11　find the temperature distribution

throughout each of the shaded regions resulting from the temperatures on

the boundaries.

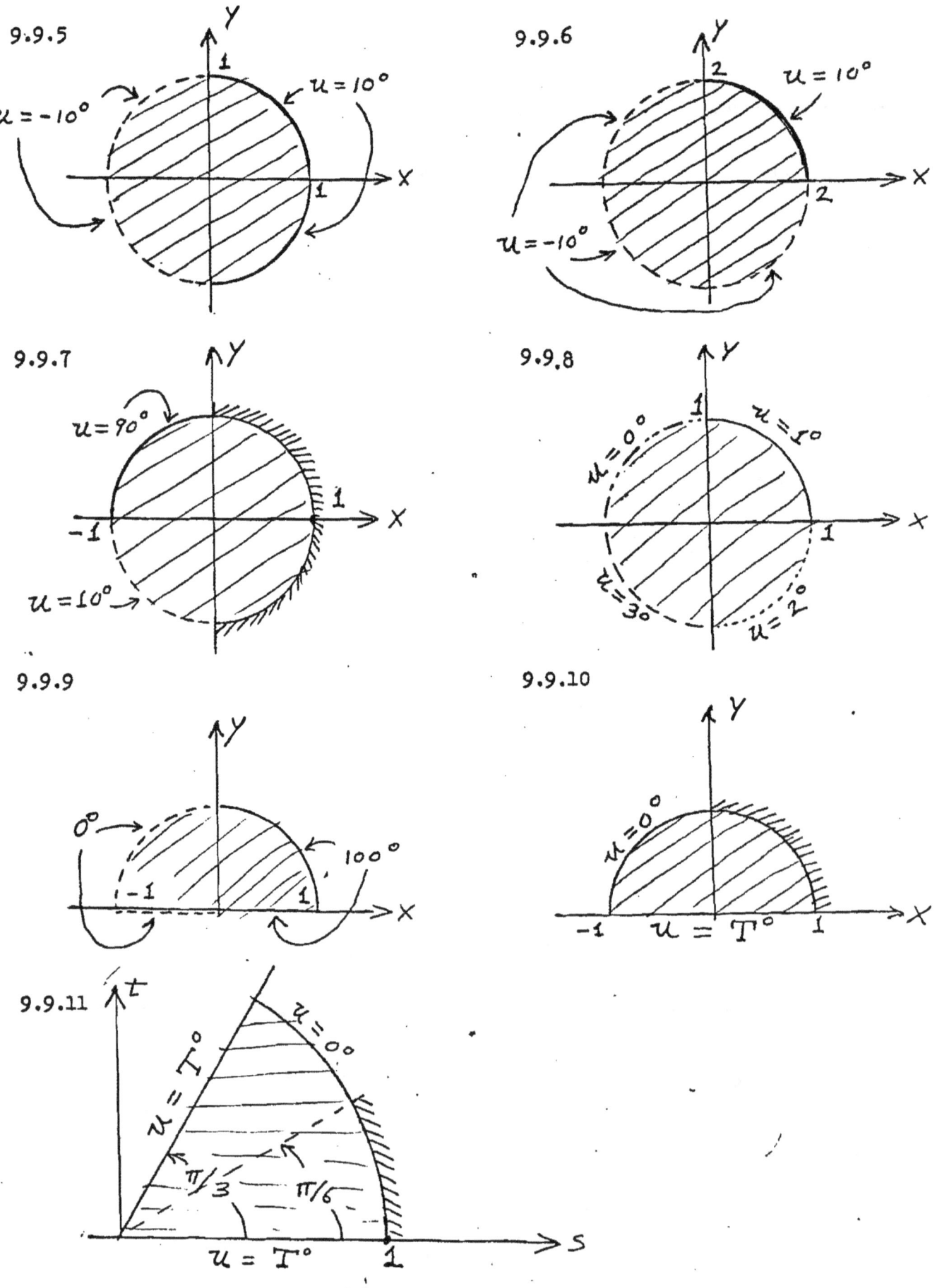

9.9.5 $u=-10°$ $u=10°$

9.9.6 $u=10°$ $u=-10°$

9.9.7 $u=90°$ $u=10°$

9.9.8 $u=0°$ $u=1°$ $u=3°$ $u=2$

9.9.9 $0°$ $100°$

9.9.10 $u=0°$ $u=T°$

9.9.11 $u=T°$ $u=0°$ $\pi/3$ $\pi/6$ $u=T°$

9.9.12 In this problem we prove that the inversion transformation

$w = 1/z$ preserves circles.

(a) Let A, B, C, D be <u>real</u> numbers and let $b = B + i C$. Show that the

relation $\quad A z \bar{z} + b z + \bar{b} \bar{z} + D = 0$

defines a circle with center at $z = \bar{b}$ and radius $\sqrt{|b|^2 - DA} / A$ if

$A \gtrless 0$. If $A = 0$ it is a straight line (circle of infinte radius).

(b) Show that under the inversion $w = 1/z$ the above circle maps onto a

a circle in the w-plane.

9.10.1 Find the bilinear transformation mapping the three given points

in the z-plane onto the corresponding points in the w-plane.

(a)	z	w		(b)	z	w		(c)	z	w
	3	2i			0	2			1	∞
	0	2			2	∞			∞	0
	-3	-2i			4	i			-1	1

9.10.2 Let $\sqrt{3} - i$ and q be points symmetric with respect to the

circle $|z| = 6$. Find q.

9.10.3 Let $4 + 3i$ and q be points symmetric with respect to the

circle $|z - 3 - 2i| = 5$. Find q.

9.10.4   Find a bilinear transformation mapping the half plane $Re(z) \leq 0$

onto the circle $|w| \leq 3$ and the point $z = 1 + i$ onto $w = 0$.

9.10.5   Find a bilinear transformation mapping the half plane $Im(z) \geq 1$

onto the unit circle $|w| \leq 1$ and mapping the point $z = \sqrt{3} + i$ onto

$w = 0$.

9.10.6   In this problem we will derive the expression for the cross ratio

(2) of section 9.10. We wish to find the bilinear transformation that

maps the three distinct points z_1, z_2, z_3 in the z-plane onto the three

distinct corresponding points w_1, w_2, w_3 in the w-plane respectively.

(a) Show that if

$$w = \frac{az + b}{cz + d} \,,$$

then

(1) $$w - w_k = \frac{(ad - bc)(z - z_k)}{(cz + d)(cz_k + d)}$$

where $k = 1, 2, 3$.

(b) Use (1) above to show that

$$\frac{(w - w_1)(w_2 - w_3)}{(w_2 - w_1)(w - w_3)} = \frac{(z - z_1)(z_2 - z_3)}{(z_2 - z_1)(z - z_3)} \,.$$

(Hint: Start with the left side of this expression and convert it into z's using (1) four times.)

9.10.7   Prove that the cross ratio of four distinct points z_1, z_2, z_3, z_4

$$\frac{(z_4 - z_1)(z_2 - z_3)}{(z_2 - z_1)(z_4 - z_3)}$$

is real if and only if they are on a circle.

9.10.8   In this problem we will prove that a bilinear transformation maps a circle C and two points $z = p$ and $z = q$ symmetric with respect to C in the z-plane onto a circle C′ and two points $w = p′$ and $w = q′$ symmetric with respect to C′ in the w-plane. We abreviate this statement by saying "a bilinear transformation preserves circles and their symmetric points".

(a) Argue that a simple translation preserves circles and their symmetric points.

(b) Argue that a simple rotation preserves circles and their symmetric points.

(c) Show that a magnification preserves circles and their symmetric points.

(d) Show that the inversion $w = 1/z$ preserves circles and their symmetric points by examining the following:

(i) A circle with center $z = \bar{b}/A$ and radius $\sqrt{|b|^2 - DA}\,/A$ is given by

C: $A\,z\,\bar{z} + bz + \bar{b}\bar{z} + D = 0$ (where A and D are real) in the z-plane. (Recall SP9.9.12). Show that under inversion this becomes

C': $D\,w\,\bar{w} + \bar{b}w + b\bar{w} + A = 0$ which is a circle having center $w = b/D$ and radius $\sqrt{|b|^2 - DA}\,/D$.

(ii) Show that p and q are symmetric with respect to circle C if and only if $A\,p\bar{q} - bp - \bar{b}\bar{q} + D = 0$.

(iii) Show that $p' = 1/p$ and $q' = 1/q$ are symmetric with respect to circle C'.

(e) Using (a), (b), (c) and (d) above, prove that the general bilinear transformation preserves circles and their symmetric points.

9.10.9   Let C be a circle in the z-plane and let $z = p$ and $z = q$ be two points symmetric with respect to C. Let Γ be a circle passing through both p and q. Show that Γ intersects C at right angles. (Hint: Map

C onto a straight line.)

9.10.10 Let C be a circle with center at O. Let Q be a point out-

side C. Construct two

lines from Q tangent

to circle C at T and

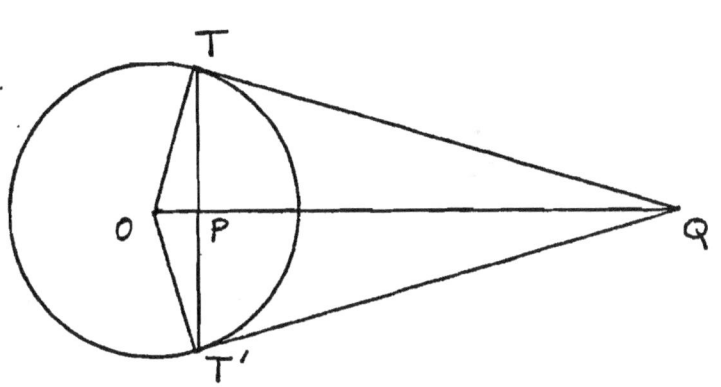

T′. Let P be the

point at which $\overline{TT'}$ intersects $\overline{OQ}$. Prove that P and Q are symmetric

with respect to circle C.

9.10.11 Let Γ_1 be the circle $|z| = 1$ and let Γ_2 be the circle

$|z + C| = R$ where C is real and positive. Assume also that Γ_1 is

inside Γ_2.

(a) Find two points, z = p and z = q, symmetric with respect to both

Γ_1 and Γ_2

answer:

$$p = \frac{1}{2C} \left[R^2 - c^2 - 1 - \sqrt{(R^2 - c^2 - 1)^2 - 4c^2} \right]$$

$$q = \frac{1}{2C} \left[R^2 - c^2 - 1 + \sqrt{(R^2 - c^2 - 1)^2 - 4c^2} \right]$$

(b) Find a bilinear transformation that maps Γ_1 and Γ_2 on the z-plane onto two circles Γ_1' and Γ_2' having their centers at the origin in the $\Im$-plane. Write the equations of Γ_1' and Γ_2'.

answers: $\quad \Im = \dfrac{z - p}{z - q}$

$$\Gamma_1' : |\Im| = p$$

$$\Gamma_2' : |\Im| = \frac{p(R - c) - p^2}{p(R - c) - 1}$$

(c) Suppose Γ_1 is maintained at temperature T_1 and Γ_2 is at temperature T_2 in the z-plane. Determine the temperature distribution $u(x, y)$ between these two circles.

Answer:

$$u(x,y) = \frac{(T_2 - T_1) \log \left\{ \dfrac{(x - p)^2 + y^2}{(px - 1)^2 + y^2} \right\}}{2 \log \left\{ \dfrac{p(R - c) - p^2}{p(R - c) - 1} \right\}} + T \ .$$

9.11.1 Use Poisson's integral formula for the circle to estimate the temperature at the point $z = -1/2$ for the disc described in problem 47 of this chapter. Use values of the temperature at the discrete points $\phi = 5°, 15°, 25°, \cdots , 355°$ as given on the boundary. Compare your result with the exact answer given by the solution to problem 47.

9.11.2 Find Poisson's integral formula for the circle $|w| \le R$. To do this, use the mapping $w = R z$ in conjunction with (1) of section 9.11.

Use the variables $w = \rho e^{i\theta}$, $z = r\ e^{i\theta}$, $z = re^{i\theta}$.

Answer: $u(\rho, \theta) = \dfrac{1}{2\pi} \displaystyle\int_0^{2\pi} \dfrac{(R^2 - \rho^2)\ U(R, \emptyset)\ d\emptyset}{R^2 - 2R\rho \cos(\theta - \emptyset) + \rho^2}$.

9.11.3   Prove that the value of a harmonic function at a point P equals

the average of its values on the boundary of any circle with center at P.

(Of course, we assume the function is harmonic at all points inside this

circle.) Hint: Use the result of the previous problem.

9.11.4   Consider the temperature distribution examined in Example 2 of

section 9.11. Use the method of that example to approximate the temperature

at $z = 2 + i$. Use the values of x_k selected in that example.

9.11.5   Get a second approximation to the temperature at $z = 2 + i$

found in the previous problem. Use the method of Example 3 with $d\theta = 10^{\circ}$.

9.11.6   Use Poisson's integral formula to find the values of a bounded

harmonic function $u(x, y)$ in $\text{Im}(z) > 0$ when the values of $u(x, 0)$ are

given on the x-axis as follows:

(a) $u(x, 0) = \begin{cases} 25 \text{ for } 1 < x \\ 0 \text{ for } x < 1 \end{cases}$.

(b) $u(x, 0) = \begin{cases} 30 \text{ for } 2 < x \\ 0 \text{ for } -2 < x < 2 \\ -30 \text{ for } x < -2 \end{cases}$.

(c) $u(x, 0) = \begin{cases} 0 \text{ for } 2n < x < 2n + 1 \\ 1 \end{cases}$ for $2n + 1 < x <$

where n is any integer.

9.11.7   In this problem we will derive Poisson's formula for the unit circle

((1) of section 9.11). Let $z = re^{i\theta}$ be a point inside $|z| = 1$. Let $\mathcal{J} =$

$e^{i\emptyset}$ be a point on the circle $|z| = 1$. Let $f(z) = u(r, \theta) + i\, v(r, \theta)$

be analytic for $|z| \leq 1$.

(a) Explain why

$$f(z) = \frac{1}{2\pi i} \oint_{|\mathcal{J}| = 1} f(\mathcal{J}) \left[\frac{1}{\mathcal{J} - z} - \frac{1}{\mathcal{J} - 1/z} \right] d\mathcal{J}$$

when $|z| < 1$.

(b) Set $\mathcal{J} = e^{i\emptyset}$ and $z = re^{i\theta}$ and show that

$$\left[\frac{1}{\mathcal{J} - z} - \frac{1}{\mathcal{J} - 1/z} \right] d\mathcal{J} = \frac{(r - 1/r)\, i e^{i(\theta + \emptyset)}\, d\emptyset}{(e^{i\emptyset} - re^{i\theta})(e^{i\emptyset} - e^{i\theta}/r)} .$$

(c) Multiply numerator and denominator of this last expression by

$re^{-i(\theta + \emptyset)}$ and simplify to get

$$\frac{i(r^2 - 1)\, d\emptyset}{2r \cos(\emptyset - \theta) - r^2 - 1}$$

(d) Combine this last result with (a) to get

$$u(r, \theta) = \frac{1}{2\pi} \int_0^{2\pi} \frac{u(1, \emptyset)(1 - r^2)\, d\emptyset}{1 + r^2 - 2r \cos(\emptyset - \theta)}$$

Without determining the constants A, B or the x_i's, write the two forms

of the Schwarz-Christoffel transformation ((2) and (3) of section 9.12)

for the mapping of $\text{Im}(z) \geq 0$ onto the polygon shown in the v-plane.

9.12.1

9.12.2

9.12.3

9.12.4

9.12.5

9.12.6

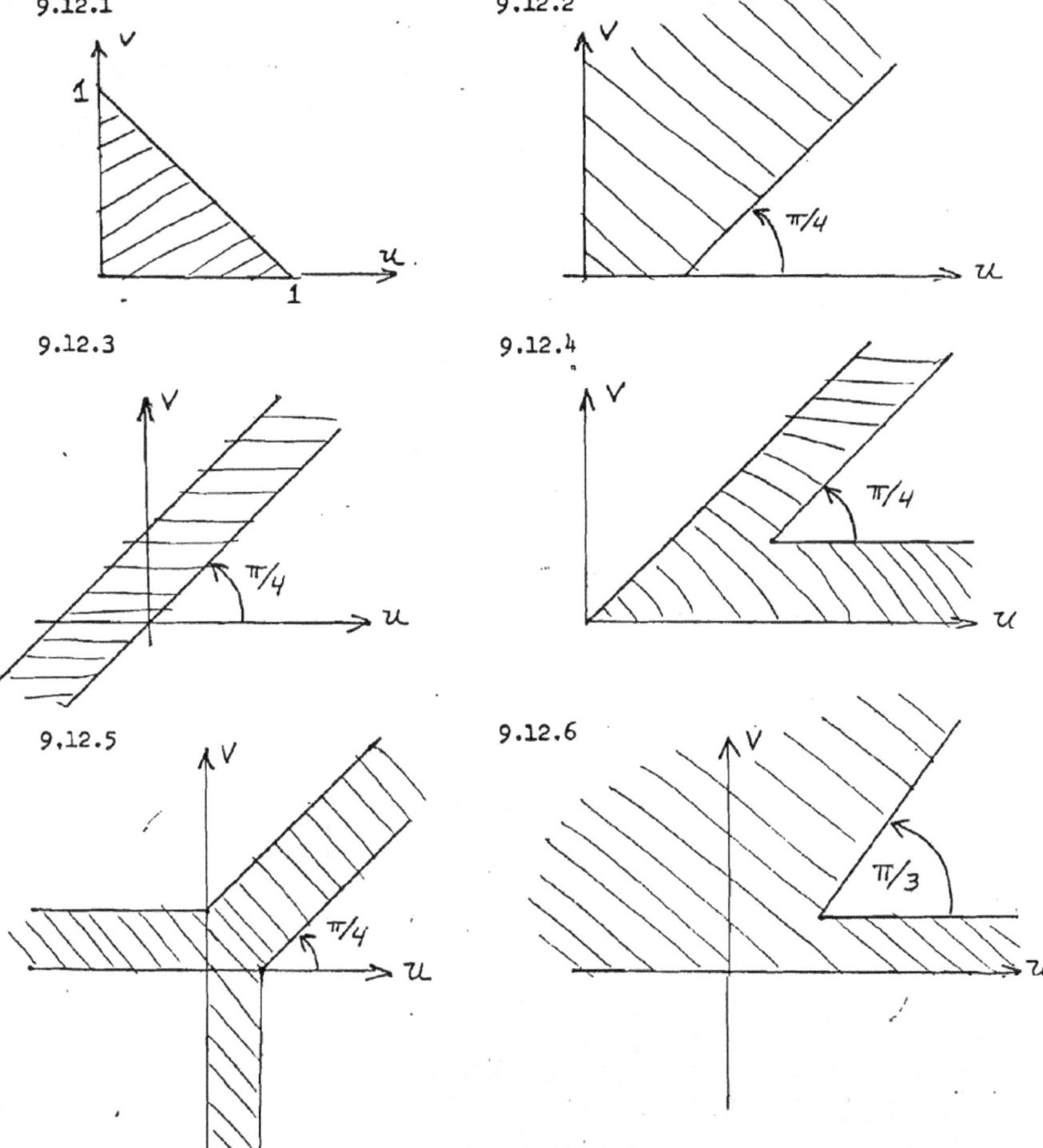

9.12.7

9.12.8

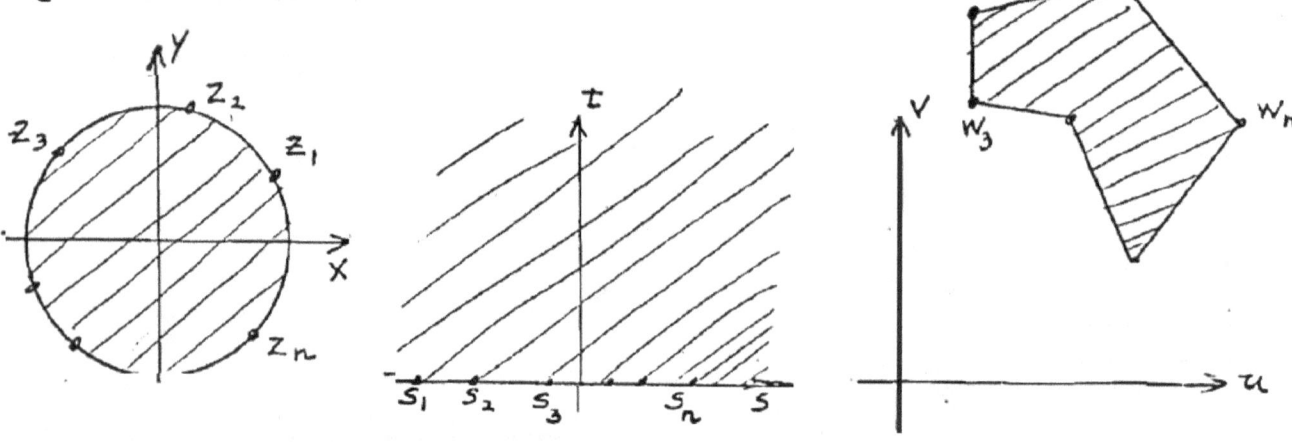

9.12.9   In this problem we will demonstrate that the function

$$w = A \int^z \prod_{i=1}^{n} (z - z_i)^{-\alpha_i} \, dz + B$$

maps the unit circle $|z| \leq 1$ onto the polygon shown in the w-plane.

The vertices w_i (i = 1, 2, $\cdots$, n) of the polygon map onto the points

z_i on the boundary of the circle.

(a) Write the Schwarz-Christoffel transformation for the mapping of the

polygon in the w-plane onto $\mathrm{Im}(\Im) \geq 0$.

(b) Show that $z = \dfrac{\Im - i}{\Im + i}$ maps $\mathrm{Im}(\Im) \geq 0$ onto $|z| \leq 1$.

(c). Using the bilinear transformation in (b) rewrite the answer for (a) in terms of z. Call z_i the image of s_i, $(i = 1, 2, \cdots, n)$.

(d) Simplify the result in (c) using the fact that $\sum \alpha_i = 2$ to get the desired mapping function.

Answers:

(a) $\quad w = A'' \displaystyle\int^{\Im} \prod_{i=1}^{n} (\Im - s_i)^{-\alpha_i} \, d\Im + B$

(c) $\quad w = A' \displaystyle\int^{\Im} \prod_{i=1}^{n} \left[\frac{z+1}{z-1} - \frac{z_i + 1}{z_i - 1} \right]^{-\alpha_i} \frac{dz}{(z-1)^2} + B.$

Use the Schwarz-Christoffel transformation to determine a function which maps $\mathrm{Im}(z) \geq 0$ onto the polygon shown in the w-plane. Determine all constants A, B and the x_i's which appear in (2) and (3) of section 9.12

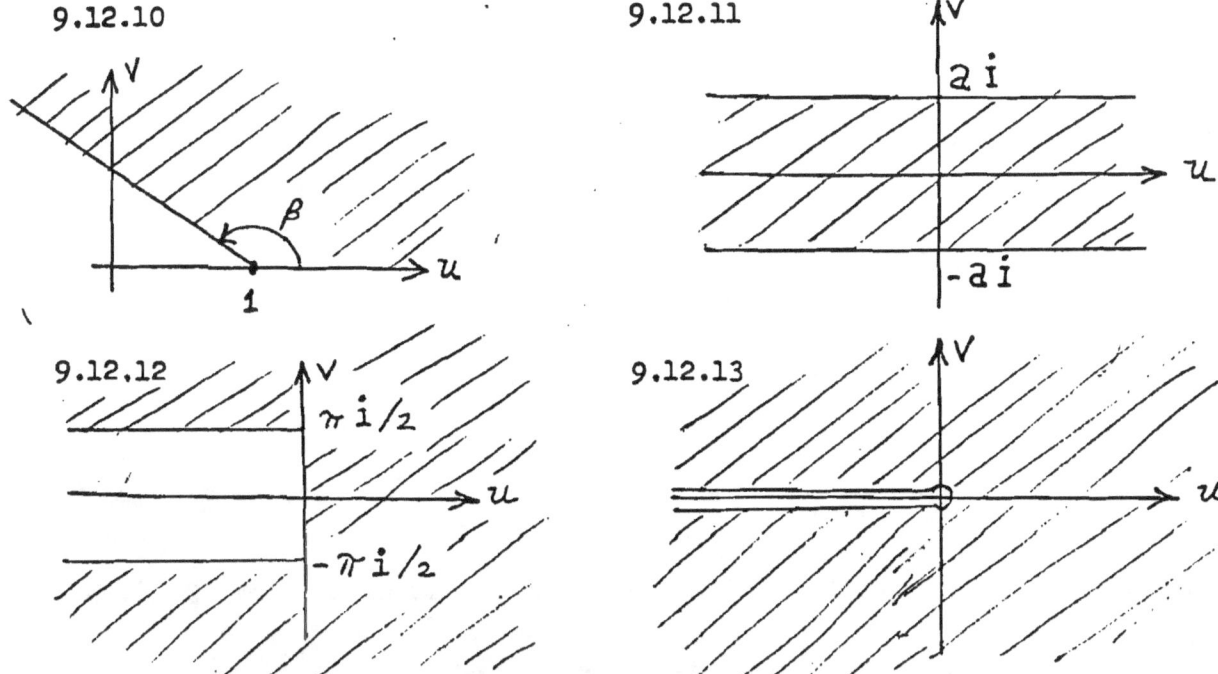

9.12.10

9.12.11

9.12.12

9.12.13

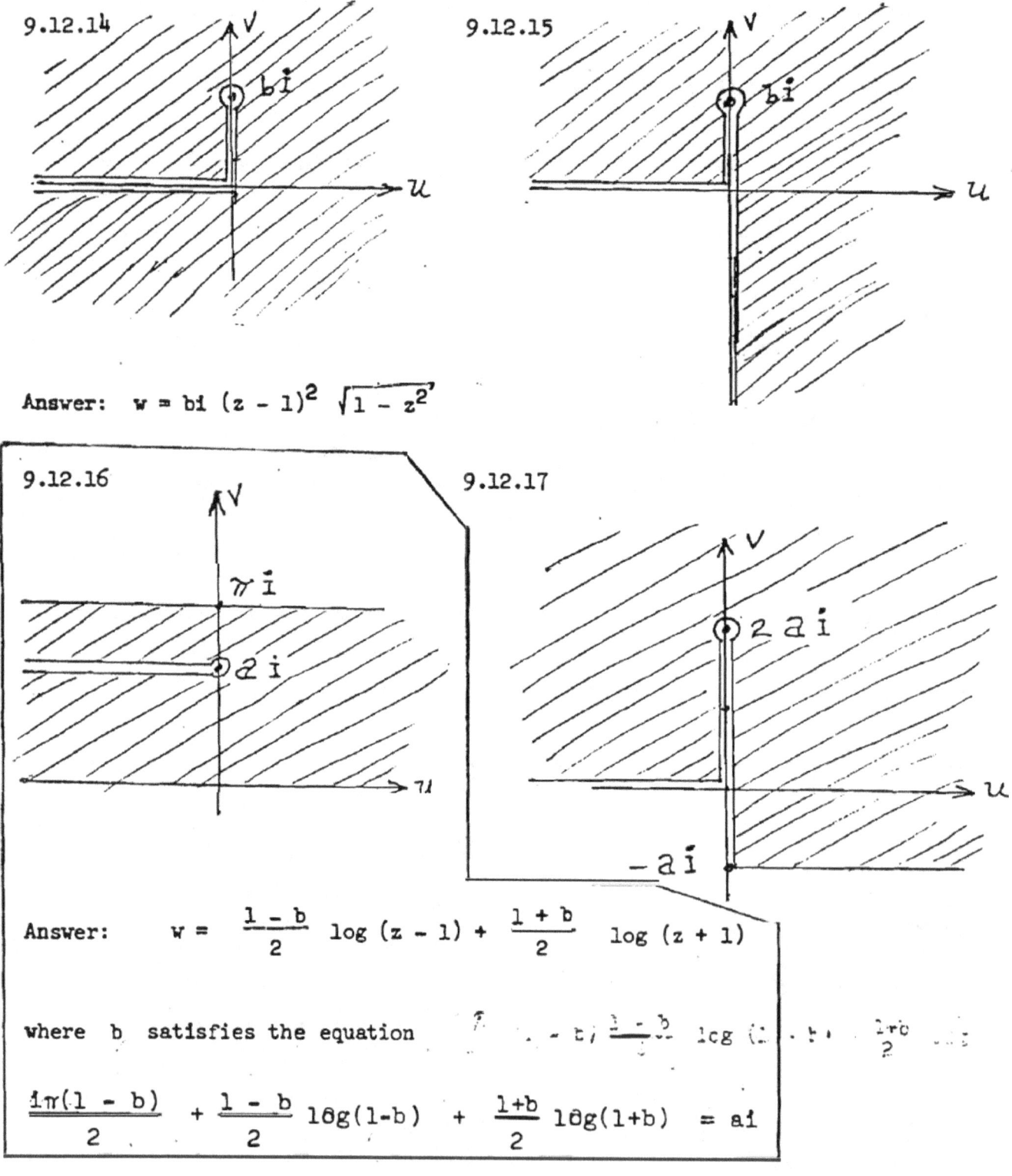

9.12.14

Answer: $w = bi\ (z - 1)^2\ \sqrt{1 - z^2}$

9.12.15

9.12.16

9.12.17

Answer: $w = \dfrac{1 - b}{2}\ \log(z - 1) + \dfrac{1 + b}{2}\ \log(z + 1)$

where b satisfies the equation

$$\frac{i\pi(1 - b)}{2} + \frac{1 - b}{2}\ \log(1-b) + \frac{1+b}{2}\ \log(1+b) = ai$$

9.12.18   Find the temperature at the point $z = 2\sqrt{2}\,i$ in the region shown

with the given boundary values. (Hint: Use the result of problem 9.12.14.

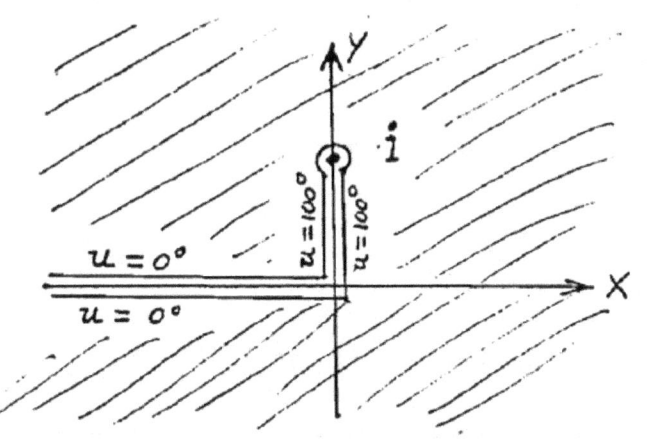

The point $z = 2 \sqrt{2}\ i$ will map

onto "i" in the upper half plane.)

Answer: 50.

9.12.19   Determine the region in the w-plane into which

$$w = \int_0^z z^{-\alpha} (1 - z)^{-\beta} dz$$

maps $\text{Im}(z) \geq 0$.   Answer:

The triangle shown where

$$\gamma = 2 - \alpha - \beta$$

and $a = \dfrac{\Gamma(1 - \alpha)\,\Gamma(1 - \beta)}{\Gamma(\gamma)}$

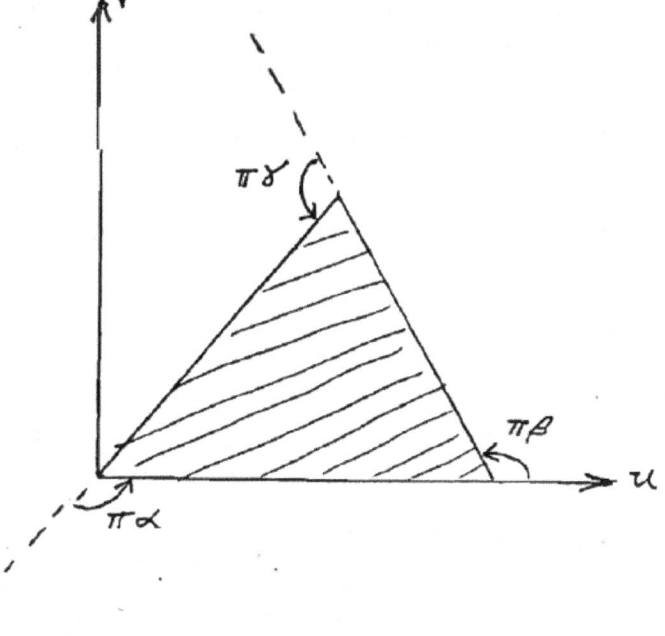

9.12.20   Show that the function

$$w = \int_0 \frac{dz}{\sqrt{(1 - z)^2 (k^2 - z^2)}}$$

with $k > 1$ maps $\text{Im}(z) \geq 0$

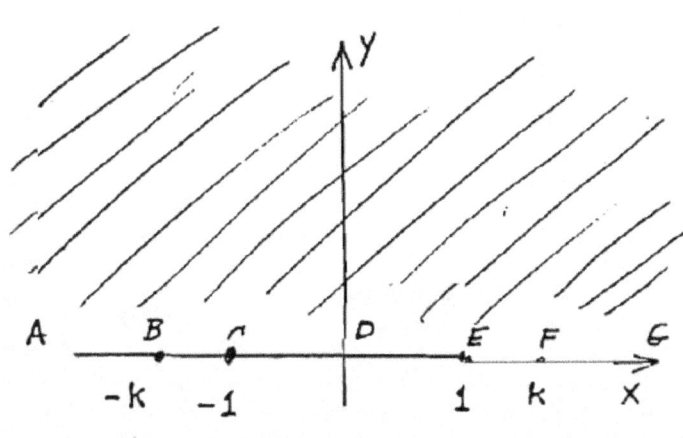

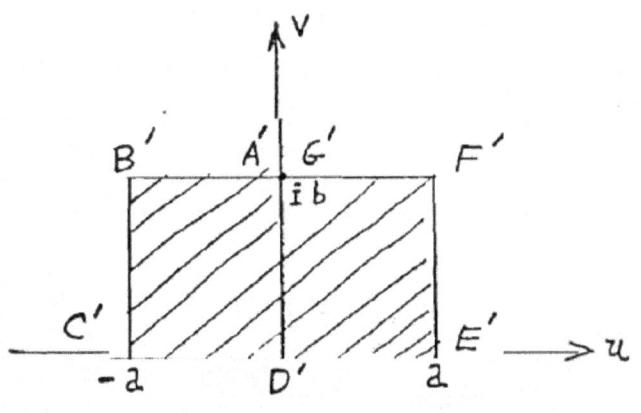

onto the rectangle shown. Show that

$$a = \frac{1}{k} \int_0^1 \frac{dx}{\sqrt{(1 - x^2)(1 - k^{-2} x^2)}} = \frac{1}{k} F\left(\frac{1}{k}\right)$$

where F denotes the so called complete elliptic integral on the left. Also

show that

$$b = \frac{1}{k} \int_1^k \frac{dx}{\sqrt{(x^2 - 1)(1 - k^{-2} x^2)}} .$$

9.12.21   In this problem we will show that the Schwarz-Christoffel trans-

formation ((2) of section 9.12) maps $\mathrm{Im}(z) \geq 0$ onto the interior of the

polygon. We use the notation employed in the first part of section 9.2

which describes the polygon, its vertices, etc.

(a) Show that

$$dw = A(z - x_1)^{-\alpha_1} (z - x_2)^{-\alpha_2} \cdots (z - x_n)^{-\alpha_n} dz.$$

(b) Show that

$$\arg(dw) = \arg A - \alpha_1 \arg(z - x_1) - \alpha_2 \arg(z - x_2)$$

$$- \cdots - \alpha_n \arg(z - x_2) + \arg \, dz.$$

(c) We will now begin

traversing the x-axis

from right to left and

we will watch the direction

of dw. Show that when

$z = x < x_1$,

arg dw = arg A - 2 π.

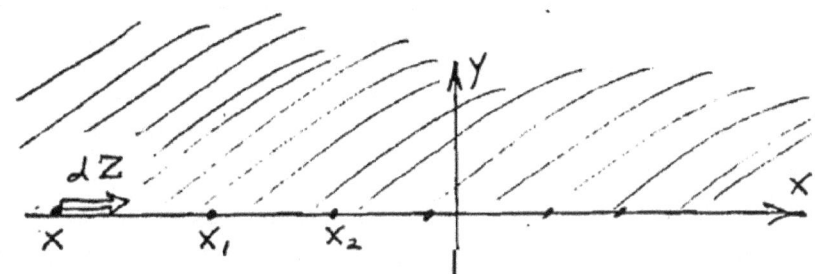

(d) Show that when x moves to a point between x_1 and x_2, arg dw

changes by $\alpha_1 \pi$.

(e) As x continues to pass over x_2, x_3, $\cdots$, describe the behavior

of arg dw.

(f) Describe the curve formed in the w-plane as z moved from $-\infty$ to $+\infty$

along the x-axis.

(g) Why does the region Im(z) > 0 map onto the interior of this polygon?

www.ingramcontent.com/pod-product-compliance
Lightning Source LLC
Chambersburg PA
CBHW080911170526
45158CB00008B/2073